Resisting Garbage

Resisting Garbage

The Politics of Waste Management in American Cities

LILY BAUM POLLANS

University of Texas Press   *Austin*

Requests for permission to reproduce material from this work should be sent to:
    Permissions
    University of Texas Press
    P.O. Box 7819
    Austin, TX 78713-7819
    utpress.utexas.edu/rp-form

♾ The paper used in this book meets the minimum requirements of
ANSI/NISO Z39.48-1992 (R1997) (Permanence of Paper).

Library of Congress Cataloging-in-Publication Data

Names: Pollans, Lily Baum, author.
Title: Resisting garbage : the politics of waste management in American cities /
    Lily Baum Pollans.
Description: First edition. | Austin : University of Texas Press, 2021. |
    Includes bibliographical references and index.
Identifiers:
    LCCN 2021007055
    ISBN 978-1-4773-2370-0 (cloth)
    ISBN 978-1-4773-2371-7 (library ebook)
    ISBN 978-1-4773-2372-4 (ebook)
Subjects: LCSH: Refuse and refuse disposal—United States. | Recycling
    (Waste, etc.)—United States. | Refuse and refuse disposal—Washington
    (State)—Seattle. | Recycling (Waste, etc.)—Washington (State)—Seattle. |
    Refuse and refuse disposal--Massachusetts—Boston. | Recycling (Waste,
    etc.)—Massachusetts—Boston.
Classification: LCC HD4483 .P645 2021 | DDC 363.72/850973—dc23
LC record available at https://lccn.loc.gov/2021007055

doi:10.7560/323700

For my parents

Contents

Abbreviations

BRC	Boston Recycling Coalition
DEQE	Department of Environmental Quality Engineering (Massachusetts)
EIR	Environmental Impact Report
EIS	Environmental Impact Statement
EPA	Environmental Protection Agency
EPR	Extended producer responsibility
ISWM	Integrated solid waste management
IWM	Integrated waste management
KAB	Keep America Beautiful
MRF	Material recovery facility (recycling plant)
MSW	Municipal solid waste
MWRA	Massachusetts Water Resources Authority
PWD	Public Works Department (Boston)
RCRA	Resource Conservation and Recovery Act
SPU	Seattle Public Utilities
WM	Waste Management, Inc.
WRWR	Weak recycling waste regime
WTE	Waste-to-energy incineration

Resisting Garbage

Introduction

I was sitting in a friend's kitchen in Seattle, Washington, in 2015. It was the first time I'd been in her sweet little home, and I'd brought her a set of candles as a house gift. I'd hastily wrapped the candles with some red cotton ribbon I'd found in my luggage. After graciously inhaling the beeswax and finding a place for the candles, she paused, looking uncertain. She was holding the ribbon in her hand, hovering anxiously over three bins in the corner. Her gaze shifted between a nearly empty trash bin, an almost full recycling bin, and the overflowing organics bucket on the counter next to her. She glanced up at me ruefully for a moment and then dropped the ribbon into the trash. "When in doubt, throw it out," she said.

My friend later explained that Seattle had introduced that phrase as part of a recent campaign to train Seattle residents to avoid contaminating their recycling and organics with nonrecyclable or noncompostable items. She told me that some of her neighbors had been fined for putting garbage into their recycling. At the time, I had never lived anywhere that even had curbside composting, let alone catchy slogans to educate residents about how to compost or recycle properly. Clearly, Seattle was serious about waste management. Boston, where I had been researching solid waste management for the past several years, barely managed to recycle. In 2015, Seattle diverted nearly 60 percent of waste generated in the city from the landfill. Boston didn't even keep track of all the waste generated within its borders; in the residential segment that it did measure, just 20 percent was diverted.[1]

Why is Seattle so serious about garbage? Why is it so different from Boston? Are the differences between the two cities' waste management practices meaningful? Is an emphasis on recycling evidence of greenwashing (a superficial effort to appear sustainable)? Or, worse, is it just an example of what sociologist Samantha MacBride calls "busyness," keeping people distracted from real environmental and social change with meaningless action? And in the context of urgent, global-scale inequality and environmental catastrophe, is garbage at the city scale even relevant? In this book I argue that waste management in our cities can differ significantly, and that the differences matter a great deal. Garbage is, after all, much more than just the stuff we throw away.

Garbage is also the very tail end of a global system of extraction, manufacturing, and consumption. Though we may think of consuming as a momentary experience—the point of purchase, unboxing a new device, eating something—it is actually just one link in a long chain. The chain begins with the extraction of a raw material from the earth, either through mining or harvesting.[2] The chain then snakes through layers of manufacturing and assembly and packaging and transportation to market, where products are purchased. These links are then followed by the inevitable: waste-making and waste-processing. On the front end of the chain, consumption demands the extraction of raw materials through mining, scraping, drilling, clearcutting, monoculture, and other environmentally devastating processes. On the back end of the chain, there is garbage. Mountains and mountains of garbage.

But waste is generated all along the chain, not just at the end, and it's generated in liquid, solid, and gaseous forms. Think tailings from mining; liquid and gaseous emissions from factories, agricultural production, and freight transport; scrap from production processes; overproduction. The last link of the chain is only the waste generated through the process of consumption itself: packaging, discarded single-use items, broken things, the things we bought and then tired of.

This trash, a very small segment of all the waste generated in our economy, is the trash we are all familiar with because it we handle it every day.[3] It is also known as garbage, rubbish, refuse, detritus, discards, or municipal solid waste (MSW). It is the part of the global extraction-manufacturing-consumption-waste chain that we all interact with daily, and that, in the United States, municipal governments control. Municipal governments decide what materials to collect and how, and they decide what to do with what they collect. Garbage is stuff we don't want; it is the product of a global chain of economic activity; and it reflects crucial material and economic interrelationships between cities and the global economy.

In this book, I compare how two American cities, Seattle and Boston, have exercised their control over garbage differently. Given garbage's multiple identities, the contrasting practices of Seattle and Boston provide lessons not only about managing trash, but also, and perhaps more importantly, about cities' agency to influence the global system of material extraction, production, and consumption. The differences in how the two cities mediate between their basic service responsibilities and the global economy show how garbage can be a lever of radical change at the local scale.

Seattle and Boston have unique histories and cultures of municipal governing and are situated within different state regulatory frameworks. They have made quite different decisions about how to manage garbage. Nevertheless, they operate within the same globalized national economy, and their citizens participate in the same globally networked consumption of the same products, made by many of the same corporations. Seattle and Boston also both operate within the same federal regulatory framework. In short, despite their differences, Seattle and Boston function within similar constraints and have similar options for waste management. As the sociologist Zsuzsa Gille has theorized, this nationally structured set of limitations and options can be understood as a "waste regime."

Introducing Waste Regimes

Gille's waste regime theory highlights the connections between national-scale economic production, individual consumption, and the generation of waste at all scales, from the individual to the economy as a whole. Because waste regime theory comprehensively synthesizes the multi-scalar identities of garbage within specific political-economic contexts, I use it as the conceptual framework for the analysis in this book. Gille's waste regimes build on Oran Young's theory of resource regimes: the collection of social, legal, and economic institutions that determine which natural resources have value, how that value is allocated, and how conflicts should be resolved.[4] Like resource regimes, waste regimes are constituted in part through policy and in part through complex negotiations informed by state and corporate interests.

In Gille's formulation, waste regimes describe the key features of how a society generates, defines, and manages its wastes:

> What appears to be unique in different time periods and different societies are the types of wastes produced (their material composition); the key sources of waste production (for example, unutilized surplus of insufficient inputs) and the dominant mode of waste circulation and metamorphosis; the socially and culturally determined ways of misperceiving waste's materiality; the ways in which, as a result, waste tends to "bite back"; the cultural, politi-

cal, and moral inclination to resolve waste's liminality (inscribed negativity or positivity); and, finally, key struggles around waste (in the sphere of production or in the sphere of distribution).[5]

In short, waste regimes "differ from each other according to the production, representation, and politics of waste."[6] In the context of waste regimes, "production" refers to how and where waste is generated within an economy.[7] "Representation" refers to how waste is defined in discourse, policy, and action. The "politics" of waste refers to who has the power to define what waste is, where it goes, and what is done with it. These are not entirely discrete categories, and it is the relationship between them that constitutes the structure of a waste regime. The extraction-manufacturing-consumption-waste chain is a key vehicle of waste production that operates at a global scale. It is reinforced and protected by governments and a host of powerful stakeholders at multiple scales; these stakeholders—public officials, corporations, advertisers—have the power to define and represent waste in public discourse. How waste is produced and represented are political questions. The political determination of who has power affects how waste is produced and represented.

Gille argues that the production, representation, and politics of waste are anchored into practice through the definition of waste itself. She identifies three aspects of waste's definition that determine how waste is created and managed within particular regimes: spatiality, materiality, and temporality. In regards to spatiality, classifying what *is* waste from what is *not* waste is a fundamental act that usually occurs in space. Defining which things are waste implies physically separating "waste" from "not waste." Individuals enact this separation at the scale of the household (i.e., deciding what goes in the bin), and individuals, businesses, local governments, waste managers, and private waste management companies work together to enact these separations at the scale of a municipal system, which moves waste objects from "our" space to "away" space.

The second key characteristic of waste within a waste regime is materiality. Many theories of waste avoid dealing with its particular material properties, instead understanding it abstractly as a social or economic process. This abstraction is also practiced by societies themselves, and it results in the tendency to misunderstand, mischaracterize, and even "misplace" waste materials. For example, if a society defines all household waste as garbage, this is an abstraction that allows the society to dispose of inert household discards (like glass or paper) alongside hazardous or toxic household discards (like detergents, solvents, pesticides, or some pharmaceuticals). Until that society differentiates the specific properties of specific materials, it does not have the means to treat those material discards appropriately or safely. The particular ways in which societies abstract or materialize discards determine the options available for management of those materials, therefore becoming an important part of a waste regime.

The final characteristic of waste within a regime is temporality. Waste exists not only in space, but also in time, implying constant processes of metamorphosis. Building on Joel Tarr's observation that US policies to prevent pollution have often actually served to displace it, or convert it into a different form, Gille argues that a wasted thing is constantly begetting and becoming other kinds of waste.[8] Gille observed in the context of socialized production in Hungary that wasted labor became surplus material, and surplus material quickly became garbage. The prioritization of disposal technologies and specific economic imperatives that are structured through national subsidy and regulatory regimes can thus define what materials in which forms have value; these definitions also shape the material possibilities of waste matter through time. For example, a national framework that prioritizes waste-to-energy incineration creates a different life cycle and set of material transformations than a framework that prohibits or is silent on incineration.

Waste regime theory allows us to examine the myriad signals about waste and wasting in our surroundings. Gille uses waste regime theory to explore shifts in waste representation, generation, and practices in Hungary as the country transitioned from what she calls "metallic Socialism," a centralized economy characterized largely by waste from inefficient production, toward a more open economy with privatized means of production that created, defined, and managed waste differently. Through her empirical analysis, she demonstrates that waste regimes are dynamic and motley; studying them allows observers to identify "resistances to seemingly ubiquitous relations of production."[9]

The contrasting stories of Seattle and Boston provide insight into what those resistances can look like in the context of what I call America's weak recycling waste regime (WRWR). I will describe the WRWR and how it evolved in detail in chapter 1, building on the work of environmental, urban, and social historians who have tracked municipal waste management in the US context. But by way of introduction, the WRWR has been carefully organized to support the extraction-manufacturing-consumption-waste chain. It prioritizes efficient waste removal and disposal and allows for the limited recycling of just a few common packaging materials. The extraction-manufacturing-consumption-waste chain, which is just another way of describing America's globalized material economy, can be understood, in Gille's terms, as the waste production machinery of the WRWR. And the whole chain is a problem.

The Scale and Consequences of American Consumption

I refer to America's globalized material economy as a "chain" intentionally. It may be tempting to think of material progression through the economy as a cycle, invoking natural processes of decomposition and rebirth. But that is not how it works. The extraction-manufacturing-consumption-waste vehicle of the US economy is a one-way path. It depends on endless extraction and infinite disposal. Yet infinitely increasing consumption is an impossibility on a finite planet; no amount of technological innovation can dodge this fact.[10] We have long since surpassed the earth's ability to support our levels of consumption; our "overshoot" now manifests in environmental disruption at all scales, from localized pollution and habitat destruction to the global climate emergency.[11]

Individual products have become more complex and more disposable than in the past, requiring more material inputs and creating more trash. Both ends of the chain, raw material extraction and back-end material processing, are globally networked enterprises. The components of a single consumer product as simple as a disposable paper cup can travel around the globe more than once as its raw materials are extracted, processed, manufactured, distributed, consumed, and thrown away. The material components of more complex products, such as mobile phones, may circle the globe again and again and again.

As consumer products become more intensive, people across the globe consume more and more. Of everything. Over the fifty years between 1959 and 2009, humans consumed more material resources than in the previous history of humanity on earth.[12] Current trends indicate that material consumption is likely to double by 2060.[13] The growth in material consumption is a global trend driven by industrialization, capitalism, and a host of powerful ideologies and institutions. But it is also a particularly American phenomenon. In the year 2018, the US economy consumed more material per capita than almost any other country in the world. Americans consumed twice as much, in terms of materials — metals, minerals, fossil fuels, and biomass (wood, food) — as the residents of the United Kingdom, and almost five times more than the residents of Kenya.[14]

It is crucial to note that the exponential rise in consumption is not driven by population growth. The University of Michigan Center for Sustainable Systems has calculated that raw material use in the United States—not including food or fuel—rose *three times* faster than population growth between 1910 and 2014. When food and fuel are included, total material consumption increased by 57 percent between 1970 and 2000.[15] The population grew 38 percent over the same period.[16]

As material consumption has grown, the share of renewable materials—such as wood, glass, and natural textiles—has declined, from 41 percent in 1910 to 5 percent in 2014.[17] This means that our current consumption is overwhelmingly dependent on nonrenewable sources, including fossil fuels and minerals. Furthermore, there has been an "appreciable decline" in the intensity of material use: materials introduced in the economy today are doing less work, and being disposed of more rapidly, than even half a century ago.[18]

The one-way system is only cost efficient because corporate producers externalize the full costs of production. The earth's ecosystems, colonized and displaced peoples, low-wage workers, and everyone else threatened by the changing climate, along with billions of nonhuman species, pay the true costs of this tangled, globally networked system.[19] Some growth in consumption—in the form of nutrition, for instance—has arguably increased quality of (human) life. But researchers across disciplines argue that much of this rising consumption does not meaningfully contribute to health or happiness and in fact may have deleterious impacts on social and psychological well-being.[20] In addition, many of the substances that we now encounter regularly in our food, furniture, buildings, appliances, clothing, personal care products, and packaging are poorly regulated inventions of the past century, many of which have been shown to threaten human and ecosystem health.[21]

At the global scale, communities that have the least power in global systems and consume the least resources face the most acute effects of the extraction-manufacturing-consumption-waste chain, including exposure to hazardous waste, climate instability, ecological collapse, and ill-conceived adaptation planning. This unequal dynamic of risk and benefit plays out at all scales. Within the United States, Black, Latinx, and indigenous people are more exposed to pollution, noxious infrastructure, and climate risk than the white elite, even though these groups have been systematically excluded from the wealth that the structures of consumption have channeled to the white elite.[22]

The US material economy, which is to say, the whole globally networked and highly unequal extraction-manufacturing-consumption-waste chain, rests on locally organized management of garbage in cities and towns. If municipalities did not efficiently remove trash, there would be nowhere to dispose of single-use convenience items and all the things we previously purchased but broke or grew tired of. Imagine your living space filled with everything you ever bought! We would either drown in our own discards or be forced into a very different relationship with things. In the United States, the WRWR relies on the local management of garbage in Seattle and Boston and everywhere in between to ensure that this does not happen. Humble local waste management keeps the whole system afloat.

Measuring and Managing Waste in the WRWR

In the United States, managing municipal waste is no small task. Between 1960 and 2017, the amount of municipal solid waste generated in the United States rose by over 250 percent,[23] surpassing the 150 percent population increase over the same period.[24] In 2017, over 50 percent of waste generated in the United States was landfilled.[25] That means that in 2017 alone, a total of 139.6 million tons of garbage, an almost unimaginable quantity, was buried in landfills, which are essentially anaerobic vaults where even organic materials will last a very, very long time. Nonorganic materials, entombed intact, will last virtually forever.[26]

Perhaps a bright spot—though a complicated one—is that as material consumption has increased, there has been a slight increase in recycling. In 2017, approximately 25 percent of municipal solid waste generated in the United States was collected as recycling, not trash.[27] An additional 10 percent of MSW was composted, up from only 4 percent in 1990. Through these two processes, a small amount of material was reintroduced into the chain. (Recycling, though, is a costly process, wasteful in itself.[28]) Roughly 13 percent of MSW was combusted with some form of energy recovery, with the resultant ash buried in landfills.[29]

In the United States, local governments decide which material discards to collect and how to process them. The national scale estimates cited above are agglomerations of thousands of municipal decisions. Within the WRWR's superstructure of waste production, representation, and politics, municipalities have relative independence to choose among regime-sanctioned disposal options—namely, incineration, landfills, and recycling (but only of certain materials). The federal government regulates disposal infrastructure through the Resource Conservation and Recovery Act (RCRA), but leaves collection and disposal decisions in the hands of states and municipalities.[30]

State governments have different frameworks, some more aggressive than others, either encouraging or preventing waste reduction and diversion activities at the local level.[31]

The differences among city waste management systems cannot be explained by differences in state policy alone. Some cities, including Seattle, recycle and compost a tremendous quantity, and Seattle does these things even though other cities in Washington state do much less. Cities in other states, such as Boston, have historically done only the bare minimum in terms of recycling (Boston has done so even though Massachusetts has ambitious and supportive state policies).[32] Even within states, cities do not implement the waste regime identically. Evidence of Gille's "resistances" can be found in these differences.

Recycling and diversion rates offer a simple but ultimately superficial means to compare solid waste management in different cities. The data available from organizations such as the Environmental Protection Agency (EPA) and individual city governments encourage this kind of shallow comparison. But, at least in the US context, recycling and diversion numbers are incomplete and problematic. Municipalities each devise their own method of counting, and most municipalities don't count everything. Many recycling and diversion numbers, including the EPA's numbers, are estimates based on models that are out of date and widely critiqued. Ultimately, recycling and diversion numbers do not communicate much, and they are not really comparable.[33]

Recycling rates and similar metrics also do not provide any information about how a city engages with the production, representation, or politics of a waste regime. In fact, recycling rate estimates obscure the municipal institutions of waste management, reducing a complex, negotiated system into a vague figure. In order to know whether Seattle's sophisticated source-separation and public-education programs are radically or just superficially different from Boston's more conventional approach to waste management, we need better methods for comparing the two systems. I have developed the wasteways framework for this purpose.

Introducing Wasteways

The wasteways approach draws on planning theory and theories of policy processes, problem framing, expertise, and infrastructure to identify meaningful differences between municipal waste systems and how they relate to the waste regime. I use the term "wasteways" in the tradition of foodways or lifeways: a means of understanding how a particular place creates its own coherent system of infrastructure and meaning for garbage within the context of a waste regime.

A wasteway is a city-scale negotiation among citizens, public servants, and the political economic processes through which waste is produced and represented. Cities operationalize the waste regime by defining the materiality, temporality, and spatiality of garbage—not in the abstractions of the waste regime, but concretely. In the practical parlance of urban governance, cities determine what is trash, how it should be removed, and where it should be taken. In theoretical terms, a wasteway is structured through the politics of problem definition, specifically, who participates and what kind of knowledge is mobilized in the definition process. In essence, municipal waste managers publicly define *the problem of garbage* at the city scale. How a city arrives at this definition, and the definition itself, constitute the backbone of a city's wasteway.

Each city's definition of the garbage problem matters, because how a problem is defined presupposes its solutions. This idea serves as the starting point for much of the literature about socio-technical lock-in[34] and suggests that solving a problem may actually rest on redefining it in new terms. As Horst W. J. Rittel and Melvin M. Webber observed in 1973, given the complex, intractable nature of planning problems, "to find the problem is ... the same thing as finding the solution.... The process of formulating the problem and of conceiving of a solution (or re-solution) are identical, since every specification of the problem is a specification of the direction in which a treatment is considered."[35] Setting the terms of the problem is *the* consequential act in the policy-making process.

Problem definition is a process of social construction that places objective conditions in a normative and political context; it draws attention to certain issues and makes the case for public intervention in some capacity.[36] Deborah Stone argues that "conditions may come to be defined as problems through the strategic portrayal of causal stories."[37] These causal stories are not random: they portray conditions as within the realm of human control rather than as accidents or nature, and they link the problematic conditions to the specific actors or actions that caused them. Problem definitions encompass two types of responsibility: they point to the cause of the problem, and they imply or directly identify the parties responsible for the solution.[38] As people define problems, they are therefore careful to infer a solution that the problem solvers have the tools, skills, knowledge, and capacity to implement.[39] Following this thinking, when a city defines garbage as a problem, the city not only defines what garbage is, but also who is responsible for creating it, who is responsible for getting rid of it, and how it should be gotten rid of.

Because defining problems also means defining solutions, the problem definition process is an arena for expressing power and for influencing consequential outcomes in the world. It therefore matters a great deal *who* gets to be involved in the problem definition process. Individual actors, corporations, and nongovernmental organizations are constantly working to create new causal stories or to draw attention to issues that they consider problematic, and this is as true in the realm of waste management as in any other arena of public policy.

A "causal story,"[40] or a narrative that frames a public problem, will almost always be grounded in tidy, convincing nuggets of information. These "social facts" become a driving force of both problem definition and solution, proxies that stand in for complex analysis.[41] These facts are only compelling to the public if they are viewed as legitimate, or "true." The degree to which policy-makers and regular people trust facts, or believe that they are true, has a lot to do with who came up with them. This has everything to do with the legitimacy of expertise.

In the field of urban planning, expertise used to mean something very narrow and specific: highly technical knowledge and a set of similarly technical skills borrowed from military planning in the wake of World War II.[42] Policy-makers deployed this knowledge and skillset via policy analysis and comprehensive planning in order to, supposedly, depoliticize infrastructure development. Well-trained experts assessed options and justified decisions regarding solutions to problems that were already defined by the same experts or political elites.[43] In these planning and policy arenas, expertise meant knowing the world through universal, abstract truths that had been rigorously established and tested by the scientific method. This abstract, positivist form of knowledge characterized postwar urban planning and civil engineering and remains evident in the engineering approach to waste management. Sanitary engineers have a specific, focused, and bounded body of knowledge that leads engineers to frame garbage problems in terms of efficiency, optimization, and disposal capacity. Questions of waste production, reduction, or representation are considered ancillary to the specific tasks of the engineer.[44]

The engineer's modern, positivist mode of expertise dominates many policy processes, but it has been deeply contested. Thomas Kuhn first challenged the sacred premise that science was neutral, incremental, and objective.[45] Building on Kuhn's observations, many have critiqued the reductionist nature of scientific or expert approaches that simplify complex social and political processes down to a series of measurable variables that can be modeled and obscure social and political dynamics as well as power relations.[46] In the contemporary American context, both scientific and policy processes have also been fused with corporate interest; the "revolving door" between industry and regulatory bodies, the political value of corporate "expertise," and well-funded lobbying have resulted in unprecedented corporate power in the production of social facts and in policy-making, especially in highly technical and science-based policy areas such as climate change, food safety, and environmental regulation.[47] In the urban context, reliance on highly technical and often corporatized knowledge has resulted in privatized, fragmented, and unequal infrastructural landscapes.[48] These critiques have implications for municipal solid waste management; municipal systems that depend on private service providers and engineering expertise to define and solve waste problems will likely yield high-tech, capital-intensive systems favored by engineers and the private sector, leaving the public sector dependent on, and beholden to, private infrastructure. (I will pick these critiques up again in the following chapter as I trace the origins of the weak recycling waste regime).

Experts and the political elite often speak the same language, or are even the same people, and can thus work together to produce social facts, reinforcing their own structural power to define public problems and control solutions.[49] In this way, politically valued expertise translates directly to power. Participatory models of planning and policy-making recognize this fact and advocate for decision processes that engage regular people. As a counterpoint to expertise or technocratic knowledge, what "nonexperts" offer to these public planning or policy processes has variously been called lay knowledge, experiential knowledge, local knowledge,[50] ordinary knowledge,[51] sustainable knowledge,[52] citizen science,[53] specialist knowledge,[54] public knowledge,[55] community knowledge,[56] embodied knowledge,[57] tacit knowledge,[58] or indigenous knowledge,[59] to highlight just a few.

In contrast to the rational, universal, abstract, and technically sophisticated knowledge of "experts," local knowledge is contextual and specific. It emerges from personal experience and is grounded in stories heard and witnessed over lifetimes.[60] It is socially and physically embedded in a particular place,[61] community, or set of practices.[62] Local knowledge is fundamentally different from expert knowledge because it is negotiated in everyday life and is explicitly informed by personal or community values. An understanding of pollution, for example, "is not defined, described, and understood in a discrete way: rather, it is intermingled with, indeed often embedded within, other important social issues."[63] In terms of waste, households and communities determine what is useful and thus what has value through processes that are almost invisible to policy-makers.[64] At the same time, though, powerful actors work to shape these household and individual decisions through public messaging and by making select products and services available.[65] If repairing something is difficult and more expensive than replacing it, for instance, most individuals don't really have a choice. Individuals always work within the constraints established by more powerful social and political actors—that is, within the constraints of the waste regime. But city governments are also key actors here. By deciding what residents can put in the trash and what is prohibited from the trash, city-scale waste managers directly inform household decisions about value. Because residents and engineers might have very different ideas about efficiency, or what is valuable, we should expect that these groups would develop very different waste management systems.

Opening up public decisions to nonexpert participants has become a key goal of planners and progressive policy-makers who aim to deepen democracy and redistribute power.[66] Proponents of collaborative planning argue that inclusive, participatory techniques can help redistribute power and produce wiser, more implementable, and more durable policy outcomes.[67] Judith Petts has argued, for example, that nonexperts are required for accurately assessing risk in real social and political contexts.[68] Judith E. Innes and David E. Booher contend that incorporating local knowledge is essential for achieving resilient solutions to complex problems in a rapidly changing, increasingly pluralistic physical and sociopolitical context.[69] Brian Wynne demonstrated that scientists and policy-makers made terrible miscalculations about both scientific and socioeconomic consequences of nuclear fallout because they ignored the specialized (non)expertise of Cumbrian sheep farmers about their farming practices and sheep grazing behavior.[70] Nonexpert knowledge has also been demonstrated to be useful for identifying less expensive, less risky, and more preventative solutions across a range of settings.[71]

Though experts and policy-makers have traditionally balked at the idea of incorporating "nonexpert" knowledge into technical and infrastructural decision-making, it has been observed that citizens often contribute at a higher level of sophistication than experts generally assume.[72] The participation of laypeople has become essential in many environmental monitoring activities, and critically, lay perceptions of environmental problems have been shown to be as accurate as—or even more accurate than—alternative professional monitoring techniques.[73] Further, lay participants in complex technical planning processes learn quickly; while they may enter processes with limited technical capacity, they increasingly have access to information beyond what is presented to them within the context of a planning process.[74] If a process is well structured and long enough, they are able to learn what they need to know to participate meaningfully even in highly technical decisions.[75]

But while many researchers emphasize the value of lay knowledge, their discussion tends to ignore the troubling fact that in order to be integrated into a shared process, local, indigenous, or other forms of knowledge are in some instances appropriated by more powerful actors to legitimize particular courses of action that are objectionable to less powerful groups.[76] In other words, the emphasis on incorporating lay knowledge into planning runs the risk of ignoring, obscuring, or reproducing the power dynamics that protect and privilege expertise, thus continuing to protect and privilege certain voices in policy processes. When done poorly or cynically, incorporation of lay knowledge is easily reduced to "manipulation" or "therapy," undermining its radical potential.[77]

Some researchers and practitioners resist the problematic appropriation of knowledge and policy process by redefining the terms entirely. Charles E. Lindblom and David K. Cohen suggest that expert-provided data is only useful and relevant in decision-making after it has been contextualized; it is through the combination of expert knowledge and "ordinary knowledge" that "usable knowledge"—information that actually informs problem-solving in a productive way—is created.[78] Following this view, a new conception of expertise has emerged that allows "greater epistemic diversity" without collapsing into "an epistemic free-for-all."[79] In this view, expertise is "not only a message, but also, and mainly, a process."[80]

Within the expertise-as-process framework, knowledge is "understood as a learning process resulting from interactions between people in a decision-making context."[81] Inputs come from multiple sources, expert and lay alike. Expert contributors are not assumed at the outset to provide answers; rather, a process of exchange among different actors with different kinds of knowledge defines and legitimizes the status of expertise.[82] Through this process, participants' understandings of the problem they seek to solve and its context is transformed.[83] The process of building knowledge *becomes* the process of defining the problem.

Processes of knowledge coproduction are necessarily messier and more pluralistic than traditional expert- or elite-led policy processes. John Friedmann has argued that planning as a discipline is uniquely situated to synthesize and "connect forms of knowledge with forms of action in the public domain."[84] But it is clear that planning professionals do not always fulfill the role of facilitators of the coproduction of knowledge in practice.[85] Furthermore, if planners are to actively take on the role of adjudicating competing knowledge claims within a process, then there needs to be institutional space for assessing such claims and for excluding those that do not, for various reasons, stand up to scrutiny.[86] In theory, however, processes that succeed in collectively defining problems and assessing relevant knowledge claims across domains, professions, and sectors, on the one hand, and those that operate within traditional channels of power and expertise, on the other, will yield different sorts of results. Coproduction means sharing the production of social facts and sharing the definition of the problem. What is at stake, then, is a fundamental redistribution of power.

Theories of expertise and knowledge production, when viewed alongside theories of problem framing and planning processes, suggest that intentionally opening up the problem-framing process to a broad spectrum of participants will yield different results from those produced by allowing problems to be defined solely by "experts" and policy elites. My conceptualization of wasteways builds on this theory: how a city defines its garbage problem determines the nature of each municipality's relationship with the waste regime. Actors that structure and reinforce the regime have tremendous power (mostly large corporations, as will be explored in the next chapter). The regime determines how garbage is produced; it determines what we, as citizens, know and think about garbage; it determines what options are available for waste management. Given what we have seen about the scale and impacts of the global system of consumption and waste generation, the actors who actively uphold the global consumption system benefit tremendously from it—at the expense of most other living things on this planet.

As city governments make decisions about how to manage garbage locally, they can privilege the voices, values, and knowledge of powerful regime actors. In the WRWR, these actors, as we will see in the following chapter, largely represent corporate producer and corporate disposal interests. When cities rely on the expertise of these groups, they reinforce regime definitions and processes, which in the WRWR means ensuring that city governments efficiently remove waste to make space for new consumption, and that they recycle just enough to soothe consumer consciences. It means that cities maintain emphasis on end-of-pipe solutions, prioritize disposal, and do not interfere with other aspects of the production chain. This kind of municipal system performs a *compliant wasteway*; it behaves as the regime prefers. The case of Boston will demonstrate what this approach can look like in practice.

But, as Gille suggests, it is possible to resist the "seemingly ubiquitous relations of production," and Seattle will show how municipal policy can do this. Cities can actively implicate producers, they can limit consumption, and they can manage wastes in ways that illuminate and threaten patterns of infinitely increasing consumption and disposability. City-scale waste managers can incorporate outside values and knowledge, define problems in their own terms, and establish their own priorities. Decisions and practices built on contrary definitions and alternative values resist the regime and yield a *defiant wasteway*. Seattle did exactly this. Some cities may do a bit of both, and these modes of acquiescence or resistance will differ depending on each city's unique historical, infrastructural, economic, and governance context.

The differing relationships between municipalities and the waste regime — wasteways — highlight not just superficial programmatic differences between municipal waste systems, such as recycling rates, but also fundamental, institutional distinctions among urban societies' approaches to waste management. If we accept that massive challenges, like the climate crisis, inequality, or environmental injustice, are symptoms of current capitalist and colonial socioenvironmental relations, then the wasteways framework arguably gets us closer to understanding the agency that cities have to mitigate or reproduce those relations. Superficial measurements such as recycling rates actually reinforce waste regime interests by highlighting *only* the activities that are sanctioned by waste and manufacturing industries. The wasteways framework allows us to ask bigger questions and integrates a broader set of actors, interests, and values into the analysis of waste systems.

To analyze a waste system through a wasteways framework means asking a specific set of questions about the process of problem definition: What is considered waste, and what is not? What are the options for management of materials that are considered waste? What kind of knowledge or expertise was mobilized to make that determination? What is the role of citizens in the process of problem definition, and what is their role in the waste management process itself? The WRWR defines waste broadly at the national scale, but how cities define waste as a problem to be solved by the public sector determines how they enact the WRWR in practice.

In the following chapters, I will explore how the WRWR came to be and what it represents, and then I will delve into the cases of Seattle and Boston to illustrate the emergence and trajectories of two very different wasteways. By following how each city defined its waste and its waste problem, and tracing the processes and modes of expertise that were privileged in its decisions, we shall see how Boston's and Seattle's divergent wasteways developed and how the contrasting approaches in the two cities affected the formation and maintenance of the WRWR.

Chapter 1 presents a historical overview of the development and evolution of waste regimes in the United States. The chapter emphasizes how
the WRWR was coaxed into being as postwar production challenged
nineteenth-century "Sanitary City" waste representations, politics, and
solutions. The WRWR was crafted by a coalition of manufacturing interests with the support of sanitary engineers, corporate waste managers, and
a sympathetic federal regulatory framework.

Chapters 2 and 3 dive into the details of how Boston and Seattle's wasteways emerged alongside the WRWR and engaged with it. I selected Seattle
and Boston as cases because the cities have somewhat similarly sized populations and economies, but very different waste management profiles. Seattle,
recognized as a global leader in sustainable waste management, has among
the highest diversion rates (the amount of discarded material that does not
end up in a landfill) in the United States.[87] Boston has been, historically,
perfectly ordinary. Until 2018, Boston barely tracked recycling and disposal
data, and it did not make that data easily available. By the 2010s, the city
was recycling barely 20 percent of its residential waste, and it did not track
or publish data on the commercial sector.[88]

Until the mid-1970s, though, Boston and Seattle's waste management
profiles were fairly similar. Neither city offered formal recycling programs.
Both cities disposed of waste in municipally owned infrastructure—Seattle
in landfills, Boston in an incinerator. But by the end of the 1970s, disposal
capacity in both cities had shrunk considerably, and both cities faced rapidly
escalating disposal costs. Both cities turned initially to local waste-to-
energy incineration as the solution. But as the cities launched planning and
decision-making processes, Boston and Seattle diverged into distinct wasteways. Seattle resisted the regime's demands, while Boston acquiesced.

Chapter 2 traces how the interplay of politics, competing infrastructural
priorities, and a narrow discourse about waste planning in Boston left underlying assumptions about garbage and its governance intact. Although the
incinerator proposal was eventually abandoned, the decision was not driven
by proactive waste planning but by other priorities. By narrowly defining
waste and waste management throughout the crisis, Boston constructed a
compliant wasteway. The institutions and practices of waste management in
the city were defined by, and reproduced, the weak recycling waste regime.

Chapter 3 then examines how the regional politics of waste disposal and
the inclusion of a wide variety of voices triggered a planning process that
led to the redefinition of garbage in Seattle. As a result, the city was able to
consider a whole new set of possible solutions that resisted key dictates of
the WRWR. In the context of this public debate, incineration ceased to be

a legitimate solution. This was the beginning of a defiant wasteway through which Seattle created a system that resists the national and global consumption and waste trends promoted by the WRWR.

By the end of the 1980s, Boston and Seattle had weathered the worst of their crises. Boston was almost exactly where it had been a decade earlier: exporting all of its garbage to regional facilities. Seattle, in contrast, was poised for massive transformation. A series of plans and resolutions had resulted in concrete changes to waste management in the city. These changes included a key shift in system goals from efficient collection to minimized disposal; the repositioning of garbage as a potential resource; and redefined roles for citizens and the state in the project of waste management.

Chapter 4 compares Seattle's and Boston's wasteways through the trajectory of their waste systems over the 1990s and 2000s. It traces how, through the defiant wasteway, continuous incremental changes in Seattle's system advanced the new vision developed in the 1980s. The result is a system whose institutions, infrastructure, and programmatic profile, in addition to its outcomes, resist regime pressure to keep waste out of sight and out of mind, even in the face of programmatic failure. Boston's approach, meanwhile, remained almost static over the next decades. Programmatic innovation did not serve a larger vision and failed to alter institutions, practices, or outcomes in any appreciable way. The chapter concludes with a discussion about the implications of the defiant wasteway as a means of resisting garbage — that is, resisting the entire global system of garbage production.

In the final chapter, I explore the practical limits of the defiant wasteway and discuss the concept of wasteways as an analytical tool. I then revisit the weak recycling waste regime in the context of resistance from cities like Seattle and widespread anxiety about plastic pollution. I speculate that the WRWR has reached an inflection point. Marine plastics and the failure of global markets to adequately recycle plastic have yielded increasing media attention and dozens of books and documentary films. This interest persisted even through the disruptive haze of the coronavirus pandemic, which itself unleashed a whole new surge of single-use plastic waste into the environment.[89] States and cities have begun to regulate single-use plastics, and many consumers have organized to shift away from some single-use products. These changes are driven by two sets of actors whose roles are narrowly and specifically inscribed in the WRWR: citizens and municipal governments. As these groups redefine themselves and their responsibilities, they are, possibly, restructuring the regime once again.

CHAPTER 1

The Evolution of America's
Weak Recycling Waste Regime

In 1980, when we pick up the stories of Boston and Seattle, the United States was in the early stages of the formation of a weak recycling waste regime. The WRWR evolved from a "Sanitary City" regime, an approach to municipal infrastructure emphasizing sanitation that grew out of a period of dramatic civic reform in nineteenth-century industrial cities. With changes in production and consumption after World War II, the Sanitary City had to adapt to a new landscape of waste production. But the WRWR maintained some key elements and assumptions from the earlier regime, and these holdover values shaped both professional and popular opinion in cities across country, including Seattle and Boston.

The Sanitary City, initially named and described by the historian Martin Melosi, was the first coherent American waste regime.[1] It emerged from the particular conditions of rapidly industrializing cities. Prior to industrialization, waste was typically managed by individuals and small collectives without formal intervention from the state. Industrial cities were dense, and big, and disconnected from natural systems of waste absorption. Traditional modes of waste disposal—for example, bringing human waste to nearby farms, feeding food scraps to livestock, or collecting specific discards, such as tallow or rags, and using them to manufacture new products—were insufficient in an industrial city context. As garbage became more visible and disease more rampant, it was publicly defined as a problem of aesthetics and public health.

Women reformers were among the first to actively represent garbage in cities as an aesthetic problem, and this perspective created the foundation for the new national waste regime.[2] The aesthetic-problem frame was widely adopted by city politicians and urban professionals. Indeed, cleanliness became an essential component of spatial order in the City Beautiful movement, whose practitioners believed it would confer social order on chaotic, poor, uneducated, and immigrant urban populations. Many local municipal art and improvement societies advocated for municipal cleanliness alongside civic art.[3]

Emerging fears about disease arose alongside concerns about the unsightly nature of open dumping. Melosi has shown how "the age of sanitation" emerged in response to Edwin Chadwick's widely disseminated 1842 *Report on the Sanitary Condition of the Labouring Population in Great Britain*, which argued that urban filth was responsible for a great deal of disease and poor living conditions. By the 1890s, municipal garbage was fully imbricated within wider anxieties about filth and miasma, and protecting public health became a core objective of the nascent Sanitary City regime.[4]

Civic reformers in the public health and City Beautiful traditions garnered media attention for sanitation issues and organized local interests to pressure municipal authorities to address refuse management. Reformers were also interested in cleaning up dirty politics—visual order represented political order as well as social order. In terms of waste management, this meant that contracting for sanitation services was not adequate. Reformers demanded that municipalities provide sanitation services directly. Civic reformers represented waste management as a symbol of a well-run city.[5]

The earliest municipal waste programs under the Sanitary City regime were organized by city health departments, but as the filth theory of disease gave way to germ theory in the late nineteenth century, people began to see urban sanitation as the best means to prevent the spread of contagious disease, and newly minted sanitary engineers began to assume responsibility for solid waste removal. In terms of their technical and scientific orientation, they were direct descendants of the early sanitation reformers, and they successfully marketed themselves, in the context of concern about corruption, as neutral experts "above the din of local politics."[6] Engineers represented waste as an apolitical problem, thus reinforcing and expanding their power to assume control over its management. With the support of reformers and politicians, sanitary engineers developed new systems to efficiently remove waste from urban centers and dispose of it in nearby facilities. By the turn of the twentieth century, sanitary engineers, "chief among the technocrat elite," had become essential figures in municipal waste management.[7]

As sanitary engineers rose in prominence, they replaced civic reformers as the primary representers of garbage in the public realm. They also became more effective marketers for their own expertise, edging out other claims of relevant knowledge. Sanitary engineers adopted the aesthetic and public-health problem definitions from reformers and institutionalized them through a professional preference for highly engineered, technologically advanced waste collection and disposal methods. Engineers eschewed open dumping and burning in favor of constructed landfills and engineered combustion.[8]

Over the late nineteenth and early twentieth centuries, engineers redefined garbage as a material. In the early years of municipal waste management, what we think of today simply as "trash" was actually several different categories of materials. For instance, Boston classified ashes, offal, combustible waste and rubbish, market refuse, street cleanings, and cesspool and catch basin cleanings as separate materials, each of which required a different collection and disposal technique. The state-of-the-art cities at the time organized separate collections so that certain materials could be profitably recycled and others could be efficiently combusted.[9] Households were responsible for keeping these types of waste separate, reusing themselves anything they could, and ensuring that they delivered anything they couldn't repurpose to the right collector. Over time, though, efficiency-seeking engineers deemed these recovery processes uneconomical and dispensed with them. Municipal waste managers combined waste categories until there remained only one: trash.

The coordinated public representations of garbage as aesthetic and public health problems, and dirty streets as a symbol of political corruption, became the backbone of the first national waste regime. In referring to this regime as the Sanitary City, I am following Melosi's seminal history of waste management. Though Melosi did not theorize the national emergence of urban sanitation as a waste regime, the emergence as he described it comprised the key characteristics of a waste regime. Within the Sanitary City regime, there was consistent production, representation, and politics of garbage. The problems of garbage as represented by reformers and engineers were all rooted in waste's visible presence in public space. The obvious solution, therefore, was municipally managed removal of waste from central cities. Engineers devised efficient methods of collection and disposal, thus delineating the waste management options available to any city government that wanted to create a visually ordered environment, protect against disease, and showcase functional service provision.

Engineers, the politicians they worked for, and the reformers who supported them together constituted and reinforced the nascent Sanitary City waste regime. They defined garbage and developed detailed expectations for how it should be handled, who should handle it, and where it belonged. In so doing, they constructed a new, nationwide understanding of the materiality, temporality, and spatiality of garbage.

The process of regime formation reshaped everyday practices of cleanliness, thrift, and wasting in American households. Historian Susan Strasser has paid particular attention to the evolution of household practices of reuse, remaking, repurposing, and recycling toward disposability. Ultimately, she argues, "disposal has been disengaged from whatever is left of household production and assigned to the technocrats who oversee sewers and sanitary landfills."[10] As sanitary engineers made it easier to throw everything away, Americans lost expertise in reusing materials and self-provisioning.

Within the Sanitary City, various groups of actors were assigned specific responsibility vis-à-vis defining and then resolving the waste problems of the age. Citizens were once the primary managers of their own discards. But in the Sanitary City waste regime, individuals played only a limited role in defining waste and managing it. Sanitary City proponents defined urban residents as victims of waste. Reformers and engineers sought to protect citizens, in no small part because they were the labor force for industry. Beyond their roles as laborers and potential vectors of disease, citizens had no active role in framing waste problems or implementing solutions. In fact, the institutionalization of waste management as a municipal responsibility effectively othered waste—and it should be noted, waste workers, who were contaminated by their close contact with refuse[11]—thereby discouraging contact between individuals and their discards.

The Sanitary City waste regime established definitions and norms around garbage that remained intact for generations and that continue to infuse our cultural understandings of waste and wasting. But as Zsuzsa Gille observed, waste regimes are constantly being negotiated, and production, politics, and representations of waste under the Sanitary City regime were not static. Perhaps the most extreme destabilization of the Sanitary City regime occurred as a result of the changes in material availability and consumption sparked by the Great Depression and then World War II. These disruptive events altered American consumption patterns and waste composition and introduced new actors into regime maintenance.

During World War II, the military and other branches of the federal government, supported by simultaneous state and local efforts, launched campaigns valorizing the collection of scrap paper, metal, and even cooking grease as contributions to the war effort.[12] These efforts effectively redefined which materials were waste and introduced the federal government as a key actor in the waste regime. Federally mandated salvage campaigns disaggregated the singular "waste stream" into its component parts, redefining many constitutive materials as valuable. The wartime redefinitions also reshaped the role of citizens: stewarding materials became everyone's patriotic duty.

Material relations quickly settled back into traditional Sanitary City logic after the war. But the context of consumption and waste production changed dramatically in the postwar period. As manufacturers adapted wartime technologies to peacetime uses, labor unions, advertisers, the government, and many other organizations central to public life instructed Americans to shop for the good of the country. Consumption was recast as a "civic responsibility."[13] For a time, the country's rapidly suburbanizing population easily absorbed the variety of new materials and products designed to fill and maintain suburban households.[14] But markets quickly saturated, and industry turned to planned obsolescence and disposability in order to maintain postwar profits.[15]

At first, as production and consumption patterns changed in the postwar era, the ideological underpinnings of the Sanitary City waste regime were renewed. Waste management was reestablished as a professional domain; all discards were once again garbage; disposal was again the goal. But mass production and the introduction of laboratory-invented convenience materials changed the materiality of garbage. At first, in popular media, disposable products were celebrated as time-saving, modern conveniences. A *Life* magazine feature from 1955, for example, explicitly positioned disposable consumer goods as a liberatory gift. The feature included a photograph of a young man, a young woman, and a small girl, all white, with huge smiles. The three are photographed from above, standing behind an overflowing wire trash can, with their arms raised merrily, as if they had just tossed into the air the plethora of trash that rains down on them. In the foreground of the image, dozens of objects shower from the top of the frame: trays, cups, straws, napkins, wrappers, ties, pie plates, a diaper, spoons, knives, empty popcorn and ice cream tubs, and what looks like a decoy duck. The headline reads, "Throwaway Living: Disposal Items Cut Down Household Chores."[16]

As the photograph shows, more consumption meant more waste, in newer and more durable material forms. *Life*'s "Throwaway Living" article mentioned disposable diapers, dog bowls, barbeque liners, and even a disposable frying pan to "eliminat[e] the scouring of pans after cooking."[17] With the rise of convenience goods, packaging waste exploded. In the eight years between 1958 and 1966, the amount of packaging sold in the United States more than doubled. Only 10 percent of this material was reused or returned to the manufacturer. The rest was put into the trash.[18]

Disposability was novel and exciting. But the rapid rise in single-use items and packaging quickly overwhelmed Sanitary City systems and gave rise to new levels of littering. As litter proliferated in public spaces, so did worry about consumption and waste in both environmentalist and popular circles. Toward the end of the 1950s, people were writing letters to the editors of the nation's newspapers, complaining about litter filling the scenic valleys of Los Angeles or lining the Colorado River.[19] Vance Packard's 1960 book *The Waste Makers*, which argued that American society was becoming defined by its waste, was a best seller.

Growing public discomfort with litter signaled some key challenges for the Sanitary City regime. The visible presence of litter indicated that the regime was not succeeding in correctly placing waste. Individuals broke out of their roles as passive consumers as they engaged fiercely in the growing public dialogue about garbage. As the production of garbage changed and grew, representations and politics of waste changed accordingly. Over the 1950s and through the 1970s, a long public negotiation took place about how to resolve the increasingly evident failure of the Sanitary City regime. The first explicit test to the regime came from Vermont.

In 1953, Vermont banned the sale of beer in disposable containers. State regulators were stunned by the cost of cleaning roadside litter; they were also encouraged by concerned farmers who viewed the volumes of litter as a hazard to their livestock and a nuisance for their agricultural operations.[20] The ban picked up on popular and environmentalist arguments that litter was a symptom of the switch to disposable products. It was widely believed that banning the products would solve the problem.

By seeking to eliminate waste at its source, Vermont's approach was a radical departure from the solutions available within the Sanitary City waste regime and presented a challenge to the underpinnings of the new postwar consumer economy. Sensing this threat, the bottling industry, with a cohort of allied corporate interests, immediately organized to contest Vermont's new law. The result was Keep America Beautiful (KAB), a national, corporate-funded nonprofit organization devoted to litter reduction.[21]

Environmental and corporate actors shared the goal of litter reduction, but they did not view "the garbage problem" in the same terms. Environmentalists and Vermont farmers placed responsibility on industry; KAB argued that the real culprits were irresponsible consumers.[22] David F. Beard, the head of KAB in the 1960s, argued that "the bad habits of littering can be changed only by making all citizens aware of their responsibilities to keep our public places as clean as they do their own homes."[23]

KAB's messaging and political organizing was successful in stamping out attempts to regulate the production and sale of disposable consumer goods. Vermont was the first and only state to ban disposable bottles. When the initial law expired in 1957, extensive lobbying by KAB and the beer indus-try ensured that it was not renewed.[24] KAB and its allies were so successful in both lobbying and controlling the public narrative that it would be more than half a century before any government entity again attempted to regu-late the production or consumption of disposable products.

Despite KAB's lobbying, public discomfort with disposability and con-sumption remained widespread. By the 1960s, solid waste registered as a concern alongside growing awareness of the toxicity of many of the new postwar compounds that were increasingly present in the environment, harming both wildlife and people.[25] Citizens and municipal governments had started to notice harmful runoff from unlined landfills and to worry about emissions from poorly maintained incinerators.[26]

In response to growing public concern, Congress passed the first national solid waste law in 1965. The Solid Waste Disposal Act (SWDA), initially under the purview of the US Public Health Service (USPHS), focused on encouraging innovation in disposal technology and increasing capacity for waste planning at the state level. The act spurred a handful of demonstration projects and encouraged states to form agencies responsible for solid waste planning. But although the act was significant as the first federal framework for municipal waste, it was incomplete in its assessment of solid waste as an issue and limited strictly to disposal.[27] The SWDA remained neatly within Sanitary City boundaries and respected the regulatory limits established by KAB: the act understood waste as a problem best managed by removal and disposal.

Unsurprisingly, given its emphasis on disposal systems, the act did little to stem the growing torrent of garbage. In the decades following its pas-sage, waste volumes in the United States soared. With packaging becoming ever more complex and substantial, every purchase added more to the pile. Moreover, the composition of waste was becoming more varied and durable as new materials, plastics in particular, entered the waste stream. Increased volume led to increased waste management costs; suburbanization and sprawl led to increased collection costs; environmental regulation led to in-creased disposal costs.[28] Waste in the postwar era, as Melosi observed, can be characterized by growth in every dimension.[29]

To give a sense of scale: the EPA estimates that solid waste generation per capita grew by 30 percent between 1960 and 1988, to over 3.5 pounds per person per day.[30] Waste composition also changed dramatically as volumes swelled. Instead of mostly organic material, trash cans were full of war-time polymers and chemicals repurposed for suburban consumption, many of them toxic—and decidedly not biodegradable.[31] Much of the new waste was plastic. According to EPA estimates, plastics production increased at a rate of about 10 percent annually between 1960 and 1989.[32]

By the 1970s, public concern about garbage was widespread and the problem was urgent. The SWDA had stayed largely within Sanitary City boundaries and failed to curtail litter or waste generation. The continued visibility and growing volumes of trash yielded new public representations. Specifically, environmentalists, social critics, and some policy-makers began to portray garbage as a new form of pollution. In 1966, the EPA produced a film that explicitly promoted this new representation. Titled *The 3rd Pollution*, the film argued that "water pollution may flow away to bother someone else downstream, air pollution may blow away to foul the atmosphere downwind, but the 'Third Pollution,' solid waste, just piles up in staggering quantities." The film notes the irony of a technologically advanced society that can't figure out trash: "We might someday be known as the generation that stood knee-deep in garbage firing rockets at the moon."[33]

The EPA was not alone in characterizing trash as pollution. The garbage-as-pollution language was evident in popular reporting about municipal solid waste from the mid-1960s onward, and in 1971, a science writer and former Senate staffer named William Small published a book called *Third Pollution* that brought the pollution language and imagery to a broader audience. Small argued that solid waste had become "a galloping problem" for the United States.[34] Noting the tepid response of regulation and industry up to that point, he made the case that disposal-oriented solutions just move the pollution around, and that "simply to shift an environmental contaminant from one place to another cannot be tolerated."[35]

The representation of garbage as pollution amplified the critiques embedded in Vermont's beer bottle law and broadly challenged the production-consumption-waste relations of the Sanitary City regime. As a *New York Times* article, reporting on the escalating challenge of waste management in cities across the country, noted, "Experts feel that the resolution of the problems may require radical changes in people's patterns of consumption and disposal, major shifts in municipal administration, and sweeping re vision of the nation's attitude toward its environment."[36] It was a ringing critique of the whole regime.

Waste was out of control. Production had changed but management had not. And in that context, the pollution language was both a representational and a political tool. Pollution, like waste, is a constructed concept; many of the substances and materials that we now take for granted as pollution were once considered neutral, or even salutary.[37] What is viewed from one angle as a symbol of progress looks from another much like a death knell. "There is no such thing as absolute dirt," observed the anthropologist Mary Douglas. "It exists in the eye of the beholder."[38] The constructed nature of what constitutes pollution means that more powerful actors in a society have an outsized role in determining what counts as harm. From DDT to lead to cigarettes to greenhouse gases, when scientists and citizen observers raised alarm, industries marshaled lawyers and lobbyists, often using tropes of uncertainty and complexity to refute, reframe, distract, obscure, and ultimately delay or prevent regulatory action.[39]

Once a substance or a process is widely considered to be harmful, or, in other words, has come to be considered pollution, then comes the task of managing it. Douglas argued that attempts to eliminate dirt were "not negative movement[s] but a positive effort to organize the environment."[40] This is a particularly useful way of understanding waste management within the Sanitary City waste regime, given that the whole regime was structured to create the appearance of cleanliness through the removal of waste. The regime organized space through the placement of garbage. As the environmental movement increasingly promoted the concept of garbage as pollution, its placement, or rather, its *mis*placement, in space became even more significant. There are a couple of reasons for this.

First, defining something as pollution means understanding it in relation to an environment.[41] Thus, the process of constructing pollution also requires the construction of purity, or a "clean" environment. In the United States, questions of purity, nature, and cities have been particularly fraught. Nature has been understood in the popular imagination as both sublime and a "wasteland" waiting to be made valuable through urbanization. This second American philosophy about nature—a vast wasteland waiting to be exploited—is evident in our language: to *reclaim* is to convert unproductive, wild land into economically productive use; to *improve* is to build a structure on wild land or otherwise reshape it for human use. The competing views of nature present the urban either as a source of oppression and sickness or as a thriving symbol of progress and civilization. These ideas have always existed in tension with each other, complicating attempts to discern the locus of a pure "environment" in the American context. What ties these perspectives together is the implicit belief that nature and the city are distinct ontological categories.[42]

With the rise of the waste-as-pollution problem frame, the two poles of nature-city crashed against each other: garbage, which had once been a distinctly urban problem, was suddenly imposing itself on landscapes beyond the city. New regional landfills were located in far-flung rural locations; garbage washed onto suburban and exurban beaches,[43] sewerage slopped onto scenic shores,[44] toxic leachate oozed out of municipal dumps and into urban waterways,[45] mysterious and carcinogenic compounds belched out of municipal waste incinerators and into the lungs of children,[46] and a scourge of disposable beer bottles littered America's scenic byways. As William Small observed in *Third Pollution*, "Not long ago, the city ... dump was considered an unsightly but seemingly manageable appendage. ... Today there is general recognition that solid wastes are a cancer growing on the land, awful in themselves and awful in the way they further foul the already polluted air and waters near them—a third pollution inextricably interlocked with the two that have been longer recognized as unacceptable environmental hazards."[47]

Small's disease metaphor reveals the way in which garbage violated the integrity of the "space envelopes" that kept city and nature apart.[48] The metaphor also associated solid waste with human health impacts, not just abstract or distant environmental damage. This messaging updated the dusty old public health origins of urban sanitation and contextualized them within contemporary environmental science. Unlined dumps emitted harmful pollution into the air and water that presented a danger to human health. Even though neither Small nor any other mainstream environmentalist recognized this contemporaneously, this danger was not evenly distributed. By the 1970s, various social movements had begun to shed light on racialized spatial patterns of inequality in the United States. Though race was rarely invoked explicitly in mainstream environmental discourse at the time, it was always a part of the waste management story in the US landscape.

As corporations marketed an increasing array of mass-produced goods to suburban consumers, a new racialized landscape of consumption and waste came into being. These goods were made for and marketed to suburbanites, who were almost exclusively white, as a result of explicitly racist housing policies, laws, covenants, and practices at all scales in the United States; those who remained in the cities—largely Black and immigrant—were left to manage with inadequate sanitation capacity, which reinforced narratives of dirtiness and pollution on communities of color.[49] Furthermore, as new subdivisions popped up in cornfields and "pristine" undeveloped areas, spaces surrounding preexisting disposal infrastructure were increasingly left with nonwhite populations. When new infrastructure needed to be sited, Black communities were often the first targeted—a fact recognized by civil rights activists distressed by the shift in national attention from racial injustice to environmental protection.[50] This racialized landscape of consumption, pollution, and waste management eventually became the primary battleground for the environmental justice movement in the following de-cades. The movement cohered as activists demonstrated the material and conceptual connections between space, race, and pollution.[51]

Although the articulation of environmental racism wouldn't become widespread for another decade, by the 1970s "tributary" movements—indigenous rights, anti-toxics, civil rights, mainstream environmentalism—had emerged. These movements grew stronger as their participants formed alliances, and by the 1980s they would form the "river" of the environmen-tal justice movement.[52] The waste-as-pollution narrative gained traction among different groups of activists as the health impacts of waste disposal became more apparent and as social justice movements coalesced. It became increasingly clear, as alarms rose in many different communities, that the Sanitary City regime was failing.

Of course, environmental and social activists were not alone in writing the political agenda on solid waste. Packaging manufacturers and other pro-ducers were listening carefully to public discourse and actively steering it toward their preferred solutions. William F. May, chairman of the Ameri-can Can Company, speaking to an industry conference in 1967, said, "Con-venience packaging, gentlemen, is rapidly becoming problem packaging, and we are caught right in the middle." His job, as he saw it, was to figure out how "[to] continue to satisfy consumer demand for easy-open, readily disposable, adequately protective packages with built in assurance that ir-responsible disposal by users will not result in advancing the litter and pol-lution problem."[53] As part of the solution, May formed an industry-wide group, the Material Disposal Research Council, charged with identifying or inventing packaging materials that would degrade and disappear once disposed of. May and his colleagues also had help from KAB, which had remained active since the late 1950s. The organizations produced new cam-paigns that—like May's address to his packaging-industry colleagues—appropriated environmentalist language and imagery while deftly shifting focus away from production.

KAB, in particular, successfully translated producer interests into com-pelling and influential public messaging. In their most notorious advertise-ment, first aired on television in 1971, a man with two long, dark braids, dressed in leather, beads, and feathers, canoes down a waterway to a dra-matic soundtrack of drums and strings. As he paddles, the camera pans to bits of plastic and paper in the water, and then zooms out to reveal a riverside built up with factories billowing smoke, making a direct link be-tween litter and more traditional forms of pollution. The camera cuts back to the man pulling his canoe up on a shore amid piles of garbage. A gravelly male voice tells us, "Some people have a deep abiding respect for the natural beauty that was once this country." The camera then cuts to someone throwing what looks like the detritus from a take-out meal out of a car window, and the voice continues: "And some people don't." The camera zooms in as the caricature of an indigenous man turns his face to us, revealing a big tear rolling out of his eye. The narrator intones, "People start pollution. People can stop it."

In the commercial and associated print ads, KAB explicitly appropriated the language and messages coming from social and environmental movements at the time. The ads clearly displayed litter as pollution—and purity as a pre-urbanized nature (the complexity of this imagery is only exacerbated by the fact that "nature" was infamously represented by a costumed Italian American actor, Iron Eyes Cody, playing the "Indian"). Within this tableau, the ads explicitly designated a responsible party for both creating and solving the problem: individual consumers. By shifting a simple behavior—throwing trash in the right place—individual consumers could protect (an imagined) pristine nature and solve the problem of garbage pollution. The image and concept became central to KAB's messaging for the next twenty years. The "Crying Indian" was regularly featured in flyers and marketing materials distributed to KAB members through the 1980s, explicitly invoking individual and municipal action as a solution to the problem of litter.

To be clear, although KAB's narrative did not directly implicate government, its desired approach relied heavily on public services. Responsible individual disposal depends on a comprehensive infrastructure of public trash cans, continued investment in street cleaning and waste collection, and a network of disposal facilities to receive the waste. The producers creating and profiting from the new tsunami of disposable objects were intentionally invisible in KAB's framing. By placing responsibility on individuals alone, KAB sidestepped contemporary concerns about toxicity, growing solid waste volumes, hazardous disposal techniques, and how the impacts of all of this might fall unevenly on different groups. KAB's framing obviated the need for more substantial environmental solutions while simultaneously positioning packaging producers as concerned citizens out to protect the American landscape and heritage. In short, KAB's approach to fixing the problem of garbage shrewdly externalized the costs of disposability onto individuals, communities, and the public sector.

The success of this framing (though not the advertisement described above, which aired the next year) was painfully evident in the activities surrounding the first Earth Day. Many Earth Day organizers and participants understood waste as a visual symbol of larger problems. In fact, Denis Hays, the national coordinator of Earth Day, explicitly argued that "ecology was concerned with the total system—not just the way it disposes of its garbage."[54] By the time Earth Day happened, the organizers understood it as a teaching opportunity in that it could help American citizens to connect ecology, toxicity, consumption, and civil rights. Nevertheless, in hundreds of Earth Day actions organized locally around the country, people simply cleaned up litter.[55] Though the actions were meant to be symbolic of transformative change, they just hid the evidence—à la KAB. They did not change the production patterns that created the garbage in the first place.

News analysis of Earth Day reflected this complexity. An article about the impacts of Earth Day on city policy in the *New York Times*, for example, noted that "solving the problem of cleaning up the environment has been especially difficult because people are throwing more things away." This phrasing, which may well have been influenced by the marketing messages from KAB, embraced the notion that the problem lay with individual consumption and waste habits, rather than with the production and marketing of more disposable products. The same article, however, went on to note that Earth Day activists wanted to attack the waste problem at its source by targeting manufacturers of "excessive one-way packaging." In a perfect encapsulation of the politics of waste production, city officials said they were interested in pursuing this possibility, but they had run up against resistance from industry, which claimed that it was simply meeting consumer demand and shouldn't be penalized for doing so.[56]

Walt Kelly playfully represented the same politics in Pogo and Porkypine posters originally designed for Earth Day. In the posters, a sorrowful little opossum surveys a forest floor full of trash, or looks ruefully over his shoulder as he spears litter on a stick and says, "We have met the enemy, and he is us."[57] The cartoons reinforced KAB's causal story and positioned Earth Day as a time to clean up litter caused by irresponsible people. Both the *Times* article and the Pogo cartoon demonstrate the power and dominance of the KAB framing, reinforcing the machinery of the waste regime in the face of changing conditions.

The popular mobilization around garbage didn't end with litter cleanups. In 1970, Congress replaced the SWDA with the Resources Recovery Act, a more comprehensive framework that addressed urgent air and water pollution associated with solid waste disposal and established an alternative set of priorities for solid waste management. The act prohibited open dumping, essentially requiring sanitary landfilling for all waste deposited on land. The statute, which was substantially revised as the Resource Conservation and Recovery Act (RCRA) in 1976, established programs to promote regional waste planning, resource recovery, recycling, and waste reduction.[58] In so doing, RCRA formalized messages coming from environmental activists about the dangerous increases in material consumption. Moving beyond the limited KAB framing, RCRA connected the issues of production, consumption, and waste and provided a framework for managing waste differently.

Though radical in theory, the law was implemented narrowly. The regional planning requirement was eventually folded up into state planning, leaving municipalities in most states relatively independent as the main agent for waste planning and management, as they had been since the rise of the Sanitary City movement. The waste reduction and reuse components were never fully implemented.

RCRA was not successful in promoting waste reduction, but it did effectively regionalize and privatize waste disposal infrastructure. Enforcement of Subtitle D of RCRA, which required sanitary landfills and banned open dumps, accelerated landfill closures across the United States. By 1980, the year the provisions took effect, the number of operating landfills had declined 50 percent from 1976.[59] Nationwide, landfills accommodated over 80 percent of household waste, but capacity was waning, with particular rapidity on the densely populated coasts. In 1988, the EPA anticipated that 70 percent of the nation's landfills would be closed within fifteen years,[60] and new landfills were proving difficult to site. On top of this, the Clean Air Act, initially passed in 1970, had forced the closure of many municipal incinerators, including Boston's.[61]

The nation, already primed by over a decade of intermittent urban trash crises, plunged into panic as disposal capacity dwindled in one city after another. The infamous voyages of the *Mobro* and the *Khian Sea*—the first a garbage scow that sailed from New York down the coast and around the Caribbean searching in vain for a port that would accept the metropolis's waste, and the second a cargo ship that illegally deposited a load of toxic incinerator ash from Philadelphia on a Haitian beach[62]—became symbols of a country drowning in garbage. Headlines in newspapers across the country trumpeted news of an imminent nationwide garbage crisis.[63]

It has been argued that the "crisis" had more to do with a lack of good information about landfill capacity—and a time lag between the closure of local landfills and the opening of privately managed, regional facilities—than an actual nationwide capacity shortage.[64] Private waste management, which had become an organized, largely consolidated industry with massive annual profits and lobbying capacity and significant power in local waste management planning, was able to mobilize the capital to plan and construct waste-to-energy incinerators and sanitary landfills in exurban and rural locations around the country that met the new federal standards. Over the late 1970s and early 1980s, a steady increase in landfill construction eventually met demand from anxious cities across the country.[65]

Martin Melosi has argued that the "crisis" framing, whether legitimate or not, "denies the complexity of the problem and ignores its persistence over time, failing to question whether it is chronic, recurrent or temporary."[66] The garbage crisis of the 1980s was indeed a node on a long and complex trajectory that involved a lack of good data, a rapidly changing material stream, shifting geographies of disposal capacity, competing representations of garbage, and widely debated problems and solutions. Even so, it constituted an immediate and expensive dilemma for municipal waste managers across the United States whose local landfills were closing, and whose disposal budgets were stretched to the breaking point.

The so-called garbage crisis was the tipping point after decades of stress on the Sanitary City. Waste was no longer out of sight; there seemed to be no more "away." As the crisis unfolded in public discourse, the EPA, solid waste engineers, municipal leaders, environmental activists, citizens, and corporate material producers wrestled over the appropriate representation and treatment of garbage. Environmentalists and their allies argued that the solution to garbage pollution was source reduction, recycling, and better product design.[67] This view was enshrined, at least rhetorically, in RCRA.

But environmentalists were not the only ones steering public discourse, popular perception, and public policy, or ultimately, setting the terms of a new American waste regime. KAB, corporate producers, and sanitary engineers with narrow definitions of waste management were equally, if not more, influential in the debate. Rhetorically, these parties co-opted environmentalist language and symbolism and brought it into the framework for a new waste regime. In practice, the coalition of corporate interests successfully limited municipal waste management options to disposal with limited recycling. What resulted was the system that I call the "weak recycling waste regime," or the WRWR.

The actors who had been key in shaping and maintaining the Sanitary City waste regime through the twentieth century played active and critical roles in the formation and evolution of the WRWR at the national scale. These actors included sanitary engineers, both corporate and municipal; a coalition of corporate production, packaging, and manufacturing interests; and, after World War II, the federal government. We will see the local dynamics of these forces in Seattle and Boston as the two cities charted different paths through WRWR formation. At the national scale, the WRWR came to embody a very specific set of representations and politics in the context of postwar industrial production.

RCRA formally laid the foundations for the WRWR in 1976, with its broad definitions, emphasis on planning, and encouragement of reduction, reuse, and recycling. By the time Congress passed the act, the EPA had a bustling solid waste office with dozens of staff members who were researching and providing policy guidance on a number of garbage-related issues. Reflecting the more progressive ideas of EPA staff and activist environmentalists, the act provided the foundation for new ways of thinking about the extraction-manufacturing-consumption-waste chain. In addition to providing a more aggressive framework for disposal infrastructure regulation, it also set the stage for broadly considering production, consumption, and end-of-pipe solutions as part of comprehensive waste management planning. In the summer of 1976, the EPA released a short position statement on the management of hazardous waste. EPA regulators proposed a hierarchy, which they argued was equally applicable to nonhazardous waste, that prioritized waste reduction above all. They then recommended a series of nondisposal activities, such as consolidation and trade, to reduce generation, and finally treatment and disposal only for those materials for which there was no other option.[68] But RCRA's ideal hierarchy was never implemented in practice. On the ground, the act was much more effective at reshaping disposal than at disrupting the extractive, one-way, material cycle of the increasingly global economy. The disconnect between the rhetorical emphasis on waste reduction and the practical emphasis on disposal is the central sleight of hand of the weak recycling waste regime.

The EPA continued to play an active role in regime formation and maintenance through the 1980s and 1990s. The agency developed policy directives that held broad influence over sanitary engineering and state policy priorities. One of the most influential moves of the EPA was the dissemination of the concept of "integrated waste management," or IWM, which formalized RCRA's suggestions for connecting production and end-of-pipe solutions (see figure 1.1). The EPA's IWM proposal suggested that municipalities use source reduction, recycling, incineration with energy recovery, and landfilling as complementary components of a comprehensive waste management system.

The report indicated that components should be understood hierarchically, with source reduction as the most important, then recycling, then energy recovery, and finally landfilling, but proposed that each community should assess its own context and choose the balance of elements accordingly.[69] This shift in emphasis devolved the power to prioritize to mu-

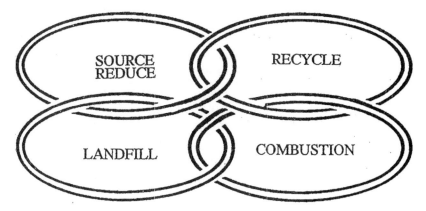

Figure 1.1. Integrated solid waste management diagram. Source: "The Solid Waste Dilemma: An Agenda for Action," Environmental Protection Agency, 1989, 17.

nicipalities, in practice leaving local waste managers without support for innovating beyond preexisting, disposal-based waste management options. Although decorated with radical language, the report served to reinforce the representations and politics of waste in the weak recycling waste regime.

The EPA proposed the integrated waste management framework to guide local planning. But it argued quite bluntly that municipalities were not alone in being responsible for solving the garbage crisis. "Who is responsible?" the report asked. The answer was straightforward: "WE ALL ARE" (capitalization in original). The EPA said that industry should have been considering disposal and recyclability in product and packaging design and that consumers should have been educating themselves about what was recyclable and choosing products accordingly; waste industry representatives should have been helping to plan and implement IWM in the communities where they worked, and officials at all levels of government should have been assisting with planning, providing guidance, and developing supportive policy frameworks.[70]

This direct call to action was a complete reframing of the Sanitary City emphasis on municipal governments—and yet the EPA did not actually manifest any of this vision in its regulations. Furthermore, while it appears to a be a shift in policy direction, it actually maintains key elements of KAB's waste framing. Specifically, the EPA document defined citizens primarily as consumers. Their role was expanded ever so slightly—they were to buy things that were recyclable, and then recycle them properly—but the EPA did nothing to ensure that recyclable products or recycling systems were available to those consumers. The limited scope of EPA action was not accidental. Various coalitions, including the bottling industry, were lobbying hard to ensure that federal regulation would not interfere with the massively profitable explosion of one-way packaging.[71]

As the packaging industry was shaping recycling policy, the disposal industry was pushing disposal technology. More and more cities experienced local disposal crunches and increased disposal costs, in part because of RCRA.[72] Disposal industry representatives responded by aggressively promoting waste-to-energy incineration (WTE). The novel technology, which was being widely deployed in Europe, combusted solid waste, generated energy, and reduced the volume of waste that required landfilling. For municipalities without space to build new landfills, WTE looked like an attractive option.

Federal programs also facilitated the development of so-called resource recovery. Energy tax credits, investment tax credits, and depreciation rules all encouraged private investment and public-private partnerships to bring WTE facilities online. Within the framework of IWM, the EPA and many states and regional planning entities actively promoted resource recovery as part of a responsible, comprehensive waste management solution. Consulting engineers, such as the firm CSI Systems, aggressively marketed incinerator technology in Boston and Seattle and everywhere in between, coaxing municipalities into investing huge amounts of capital into the facilities. The waste industry took good advantage of these programs, building facilities privately or in partnership with municipalities until financing rules were changed in 1987. The combination of federal incentives and aggressive industry marketing catapulted the new technology into use. By 1989, there were 100 modern WTE plants operating in the United States with 115 more under construction. Some anticipated that as many as 400 incinerators would be operating in the country by 1990.[73]

Ultimately, WTE did not take off as successfully as its proponents hoped. By the 1990s, public enthusiasm had cooled considerably, largely because of citizen opposition.[74] Combusting plastic and organic materials together creates compounds such as dioxin, a potent carcinogen that, once introduced, is persistent in the environment. In addition to emitting pollution, many argue that incinerators discourage recycling and waste reduction and exacerbate environmental injustice.[75] Some states, including Massachusetts, actually passed moratoriums on incinerator construction when it became clear that pollution-control issues had not been fully resolved.[76] Despite industry's best efforts to sell the technology through greenwashed labeling using terms like "energy recovery" and "resource recovery," which attempted to position WTE as a responsible and efficient form of waste management akin to recycling, the disposal technology lost momentum. Once WTE was off the table, cities struggling with reduced landfill capacity were left with recycling as the only viable end-of-pipe alternative to landfilling.

As the EPA and other federal agencies promoted WTE to the public sector, the EPA's emphasis on IWM also influenced both state solid waste planning and the sanitary engineering discipline. Engineers institutionalized the EPA's IWM vision and the emphasis on high-tech waste disposal with their own version: integrated solid waste management (ISWM). Throughout the 1990s, ISWM dominated the technical waste management literature and was institutionalized in state and local solid waste plans around the United States. Initially, ISWM emphasized increasing the efficiency and connectedness of municipal waste management processes, including source separation, collection, transportation, treatment, and disposal.[77] It also explicitly incorporated concerns about the environmental impacts of waste management. One ISWM textbook for sanitary engineers, for example, argued that society demanded more than the traditional protection of health and safety: "As well as being safe, waste management also needs to look at its wider effects on the environment."[78]

The EPA soon updated its initial ISWM diagram and began to present the "waste hierarchy" as an inverted triangle that graphically represented the proportions of the waste stream that should be handled by each technique (see figure 1.2). ISWM aimed "to minimize the quantity of waste requiring disposal and to [maximize] recovery of material and energy from waste."[79] To achieve this objective, waste professionals practicing ISWM emphasized planning and promoted a variety of techniques, including reduction, recycling, energy recovery, and disposal, to manage municipal solid waste, because "no single method of waste disposal can deal with all material in waste in an environmentally sustainable way."[80] The ISWM hierarchy explicitly prioritized energy recovery—that is, resource recovery, such as WTE—above disposal, positioning it as a form of recycling rather than a form of disposal. This positioning has long concerned environmentalists and environmental justice advocates who argue that WTE is polluting and discourages reduction and recycling. The emphasis, they say, shows the powerful presence of the disposal industry and sanitary engineers in national solid waste policy-making.[81]

Although ISWM was broadly adopted in state and local policy, engineers did not actually implement the hierarchy. Engineering textbooks on ISWM provide some clue as to why. P. R. White and colleagues, for example, argued that, although source reduction and waste minimization were positioned at the top of the hierarchy, "in reality, source reduction is a necessary *precursor*

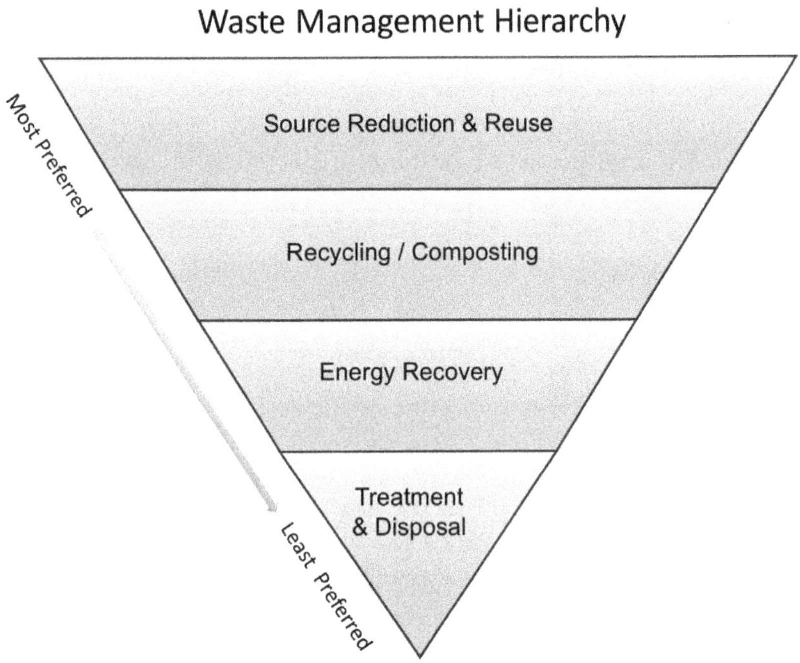

Waste Management Hierarchy

Figure 1.2. Waste management hierarchy. Source: Environmental Protection Agency.

to effective waste management, rather than a part of it. Source reduction will affect the volume, and to some extent, the nature of waste, but there will still be waste disposal. What is needed, beyond source reduction, is an effective system to manage this waste" (emphasis added).[82] In other words, while source reduction may be the most important ingredient of ISWM, it is outside the scope of sanitary engineering.

Instead of a triangle, the textbook provides a bull's-eye diagram that includes only end-of-pipe recycling and disposal options, each with equal emphasis, because each option would have different efficiency in different contexts (see figure 1.3). Municipal waste managers, the text argues, are only responsible for the material designated as waste; thus the reduction and diversion activities at the top of the hierarchy are not actually in their purview. Explicitly grounded in a rational-scientific worldview, the engineer-devised version of ISWM sought to increase the efficiency of *disposal practices* by recovering material and energy through economically viable means. Waste managers manage waste; product design, consumption, and waste reduction are ancillary concerns to be dealt with by someone else.

Even as early as the 1970s, material scientists were arguing for extending the life of consumer products, making products repairable, and eliminating planned obsolescence. Because recycling always results in degraded material, especially in the case of plastic, it therefore would not likely be a long-term solution. It was also widely understood that recycling was not economically feasible given the state of materials, technology, and collection systems. Recycling grew from 5.6 percent to 14 percent of the municipal waste stream between 1960 and 1980, but this was almost entirely individually and privately organized.[83] In this context, materials scientists, ecologists, environmentalists, and other activists felt that American society could no longer "afford to look at disposal as an end unto itself."[84]

One alternative solution did make its way into practice at the time: beverage bottle return programs. "Bottle bills," which require deposits on certain beverage containers, and then require retailers to accept those bottles back for recycling, began to spread in the early 1970s. They were less aggressive than Vermont's bottle ban, and they put the burden on retailers rather than producers. Nevertheless, beverage and bottle manufacturers and

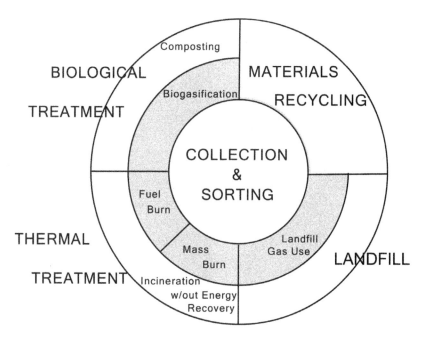

Figure 1.3. ISWM model from a popular engineering textbook. Reprinted by permission from P. R. White, M. Franke, and P. Hindle, *Integrated Solid Waste Management: A Lifecycle Inventory* (Boston: Springer, 1995).

retailers fought bottle bills hard. They were still viewed as an inappropriate infringement on industry and profit. Ten states succeeded in passing bottle bills; many others tried and failed. Massachusetts has a bottle bill; Washington State does not. But in Massachusetts, as in many other states with bottle bills, the deposit remains five cents, just as it was when the program was first established. In Massachusetts, as in states across the country, attempts to increase the deposit and expand the types of eligible bottles have consistently failed as a result of strong industry lobbying (a handful of states have managed to modestly increase bottle deposits from five cents to ten cents, far below what would be predicted by inflation). Though bottle bills have been shown to be among the most effective tools for increasing recycling, and can add substantially to state budgets when uncollected deposits are channeled to state agencies, industry opposition has ensured that they remain a relatively small piece of the country's institutional infrastructure for recycling.[85]

As it became clear that WTE was not going to be a silver bullet solution, many individuals and some municipal governments elevated municipal recycling, believing that it could alleviate some of the long-term financial and environmental costs of garbage. Recognizing this potential, corporate producers assisted in the effort in order to use recycling as a shield against more impactful environmental regulation. As Bartow Elmore has shown, Coca-Cola and other beverage companies used the promotion of government-run recycling programs to divert attention away from source-reduction legislation (like Vermont's 1953 bottle law) and to diffuse growing anxiety about wasteful consumption.[86]

Because early municipal recycling programs were developed under the influence of Coca-Cola, KAB, and the packaging industries, they focused only on the most common packaging materials. Early municipal recycling programs accepted paper and metals, and sometimes glass. More recently, municipal programs began accepting plastics—though the success of actually recycling plastics has always been limited.[87] As Samantha MacBride has shown, the choice of materials included in curbside recycling programs was not driven by economic or environmental logic—other materials, such as organics and textiles, make more sense to recycle than glass, for instance.[88] During the 1980s, the EPA promoted recycling of a broad range of materials, including paint, oil, construction materials, organics, and textiles.[89] But as municipal recycling programs became widespread, they started focusing narrowly on common consumer packaging materials: municipal recycling programs served the interests of packaging industries.[90]

As popular interest in recycling programs grew, the increasingly powerful solid waste industry made token efforts to provide recycling services to municipalities. Though the makings of a recycling industry existed in many urban markets—thriving scrap businesses, paper recycling mills, and sometimes private recyclers that could collect and process metal, glass, and paper—many municipalities simply asked the big companies with which they already had contracts, such as Waste Management, Inc., to start collecting and processing recyclable material. For decades, the big players in the waste industry made cosmetic efforts at providing recycling services while continuing to make fortunes off of landfills and incinerators.[91]

This minimal investment in municipal recycling satisfied local waste professionals who were trying to implement ISWM. It satisfied industry, whose marketing whizzes and "environmental" nonprofits could now argue more convincingly that one-way packaging was not a long-term problem, and who could use their platform to admonish consumers to put litter in the recycling bin. It satisfied the waste industry, which could offer, and charge for, a new service without much capital investment.

The only stakeholders this arrangement would not work for would be anyone actually interested in reducing the social and environmental costs of profligate production and consumption. Recycling was a fig leaf designed to make consumers and regulators complacent. This is why the environmental critic Timothy Luke, along with many others, considers American recycling to be nothing more than a "ruse."[92]

The WRWR shares some key characteristics with the Sanitary City regime, particularly in the assumption that cleanliness is paramount and waste is best managed by removal. But the weak recycling waste regime has a distinct structure, visible in the systems of waste production, representation, and politics. Waste and its production in the WRWR differ from waste and its production in the early industrial era. Waste is no longer the largely organic byproduct of household activities. Garbage is increasingly constituted by products designed to be garbage, materials that are waste before they are ever purchased.[93] The material throughput from disposable and other products with short life-spans is more durable and more toxic, with long-term implications for disposal infrastructure and the environment.

Moreover, WRWR waste representations differ from Sanitary City waste representations. Just as Zsuzsa Gille observed in the transition from "metallic Socialism" to a "chemical regime" in Hungary, the articulation of waste as pollution in the US context was an important trigger in the regime change process.[94] In the Sanitary City, waste was represented as an aesthetic and public health problem. Waste as pollution indicated that waste was dangerous, not just a nuisance, and something that caused harm to more than just the urban or human world. The waste-as-pollution framework made the relationships between cleanliness and whiteness, and dirt and nonwhiteness, conceptually accessible even as those connections were appropriated and exploited by industry. The concept of waste as pollution initially inspired a widespread belief that garbage could be prevented at its source—though this belief also was co-opted and transmuted by industry. Using the imagery and language of waste as pollution, manufacturing, beverage, and packaging interests crafted a new causal story that made people into the problem, and crafted ideological and practical solutions that would leave production interests insulated from public scrutiny.

A coalition of production and solid waste corporations then co-opted public interest in recycling, ensuring that any costs of recycling would fall on individuals, who must do the labor of separating materials; municipal governments, which coordinate local material streams; and taxpayers, who foot the bill for waste collection and treatment services. Production-side industry resisted bottle bill expansions—and any suggestion of other forms of extended producer responsibility—but promoted recycling of common packaging materials. The waste industry benefited from the new emphasis on limited recycling by offering recycling to their municipal clients, thus opening a new profit opportunity without substantially impacting the bread-and-butter business of waste disposal.

The corporate appropriation of waste-as-pollution and recycling demonstrates a key shift in the political balance from the Sanitary City regime. The WRWR sidelined reformers, city boosters, and urban professionals. City governments became tools for enacting programs that allowed both the manufacturers—garbage producers—and large-scale private disposers to profit from endless, ever-growing streams of garbage. Engineers remained paramount, but their role was hidden rather than celebrated. They were embedded in private waste management corporations, making key decisions about the technologies and maintenance of waste infrastructure. Or they were obscured within municipal waste management departments, channeling a particular and narrow knowledge base that prioritized waste disposal as a kind of legacy from the Sanitary City roots of sanitary engineering education.

In the WRWR, citizens are called "waste generators," a title that artfully conceals the actual producers of the mountains of toxic and disposable products that constitute waste. Citizens are chastised to prevent litter and to recycle, but marketing experts, industrially funded nonprofits, and some government agencies work assiduously to ensure that consumer behavior does not fundamentally change. This coalition of actors, still working through nonprofits, including KAB and the Recycling Partnership, advocate for municipal recycling and keep public attention focused on a narrow set of issues, carefully avoiding engagement with production-side regulatory approaches. In other words, as in the Sanitary City, in the WRWR citizens are expected to remain passive consumers and dutiful "thrower-awayers." Their primary role is to ensure that garbage remains invisible by hiding it in the correct place.

Within the WRWR, garbage is understood as harmful. But it remains, as it was in the Sanitary City, an abstraction. Its material characteristics are hazy; citizens are encouraged not to worry about the history or future of objects they throw away. And the infrastructure for managing waste remains, for many—though of course not all—invisible. Industry actors, on both the production and disposal sides, actively encourage people not to worry about garbage at all.[95]

Since the 1980s, the WRWR has led to widespread municipal recycling in cities and towns across the United States. The corporate producer and disposer interests have successfully sidelined attempts at upstream material management, or even expanded recycling, ensuring that limited recycling remains an option for consumers who might otherwise worry about the environmental impact of consumption. Despite the effectiveness of corporate control over the WRWR, each municipality has its own unique relationship to the regime.

Seattle and Boston both considered and rejected WTE incinerators, and they both initiated municipal recycling programs in the 1980s, following national trends as the WRWR emerged. But despite these programmatic similarities, the two cities' wasteways developed differently and then continued to diverge over the coming decades. In the differences between these wasteways, there is evidence of resistance to the WRWR. By comparing Seattle's unusual system to Boston's more conventional one, we can now start to see how cities can use waste management to manipulate and to defy the whole extraction-manufacturing-consumption-waste chain.

CHAPTER 2

Non-Planning for Garbage in Boston

Boston is a city of infrastructural innovation. Much of the city is built on land conjured from swamp.[1] Space for the nation's earliest subway was carved through the heart of downtown.[2] Boston led the nation in urban renewal, razing entire neighborhoods on the modernist promise.[3] It led the world in highway deconstruction when it buried the elevated Central Artery downtown and developed a greenway in its stead.[4] It led again in environmental remediation when it brought Boston Harbor back to life through the construction of a regional sewer treatment facility.[5]

Many of Boston's infrastructural works have been controversial and overpriced. Some, such as the urban renewal of the West End, have been regretted in retrospect.[6] Others, including City Hall, have earned as many admirers as critics.[7] And still others, notably the sewage treatment plant and the sun-drenched greenway in the footprint of the old elevated, have been successful by many measures in spite of controversy.[8] But whether successful or not, Boston's leaders have been progressive and ambitious.

The city's garbage never received the same attention as other public works. It has remained, in the words of Susan Leigh Star, "invisible, part of the background for other kinds of work."[9] Even when a regional disposal crisis in the late 1970s brought trash tumbling into public discourse, the visioning power of Boston's leaders failed to materialize. Throughout the rise and eventual resolution of the crisis, city leaders remained loyal to Sanitary City principles: their aim was always to make waste disappear as efficiently and hygienically as possible.

This chapter explores how city officials initially responded to the loss of regional garbage disposal capacity in the early 1980s. It traces the technocratic origins of the city's initial incinerator proposal and then follows the various detours the proposal took before ultimately getting rejected. The chapter also analyzes how the city's narrow waste planning process kept decision-makers laser-focused on a disposal problem while they systematically excluded alternative framings. Lacking a robust plan, the city ultimately allowed political rivalries and competing infrastructural priorities to govern the final decision about where the city's garbage should go.

The Disposal Problem

By the mid-nineteenth century, Massachusetts had developed a reputation for good public health. Both the City of Boston and the State of Massachusetts embraced the Sanitary City movement and invested in research and practices to establish a clean and salutary urban environment.[10] As early as 1905, Boston was gathering statistics on waste generation by season and neighborhood and planning service accordingly.[11] At that time, and for decades onward, Boston relied mostly on local, municipally owned dumps and incinerators for waste disposal. Over the first half of the twentieth century, the city slowly discontinued its habitual use of coastal swamp areas as landfills, converting local dumps one by one into development sites.[12]

As the city converted dump sites to other uses, local waste management capacity dwindled. In alignment with national trends of solid waste industry consolidation and infrastructure privatization, by the 1970s Boston found itself relying largely on regional disposal facilities owned by private corporations.[13]

In 1975, Suffolk Superior Court ordered Boston's only incinerator to cease operations on account of severe and chronic air quality violations. The incinerator, which the city had constructed in 1960 on an industrial site known as South Bay, had no modern pollution prevention technology and had been repeatedly cited for emissions violations in the wake of the Clean Air Act.[14] It was the only facility receiving residential waste within city boundaries, so when it closed, Boston was forced to export all of its residential waste to private regional facilities. After the South Bay incinerator closed, the city started delivering about half its waste to a transfer station in Boston's Roxbury neighborhood, which was owned by a regional waste disposal company that also owned landfills across New England. The remainder was sent directly to a brand-new, privately owned and operated incinerator located in a wetland in the suburb of Saugus, about ten miles north of Boston.[15]

The Gardner Street landfill in West Roxbury, which had received nonhazardous commercial MSW, closed in 1980. After this closure, Boston had no disposal capacity, public or private, within city limits.[16] Meanwhile, private landfill and incinerator operators across the region became increasingly anxious about their ability to site new facilities. RCRA regulations and rising local resistance increased the difficulty and cost of landfill development, and state law made it relatively easy for municipalities to deny permits. Landfill operators raised tipping fees to protect existing capacity.[17]

At the same time, many existing landfills in the state were not in compliance with state and federal laws. Dozens of municipalities in Eastern Massachusetts were using unapproved landfills that had little to no barrier for capturing leachate and protecting groundwater. Many communities in the state only began to separately collect household hazardous waste in the mid-1980s, meaning that the leachate from unlined dumps was saturated with hazardous compounds from paint, caustic cleaners, batteries, pesticides, and other toxic household materials. By the early 1980s, these materials were "biting back."[18] Leachate was contaminating groundwater and poisoning pets and wildlife, forcing landfill closure, and imposing budget-breaking cleanup costs.[19]

These were not new problems: regulators had been overlooking toxins in unlined dumps for years. Before the pollution materialized, the disposal capacity deficits were more worrisome than the potential future harms. But once the pollution's effects became impossible to ignore, the costs of cleanup compounded the lack of capacity, throwing Massachusetts's solid waste systems into crisis mode.[20]

The regional disposal constraint led to increased costs, both for the city and for the commercial sector.[21] Business leaders began to pressure both the city and the state to develop more disposal capacity to contain costs. The administration of the governor at the time, Michael Dukakis, was considering a broad recycling initiative that would reduce the overall disposal burden statewide. But to city officials, such a plan was untested and seemed irresponsible. The path forward was clear for Boston mayor Kevin White's administration: more disposal capacity.

White's public works commissioner, Joe Casazza, was a longtime public servant who worked in the tradition of the Sanitary City. His priority was clean streets, first and foremost. Casazza was closely involved with Keep America Beautiful, even joining the organization's board of directors in 1981.[22] He and his staff were involved in dozens of small programs to reduce litter, curb illegal dumping, and encourage residents to tidily set out their garbage cans for collection. The issue of cleanliness was an interest of Casazza's independently, but an emphasis on clean streets was also reinforced by business owners and residents, who frequently wrote to complain about road litter, trash-strewn lots, and sloppy garbage collection. Administration officials responded eagerly to such constituent concerns.[23]

Casazza was particularly thorough in his responses to business representatives who requested litter baskets to keep business districts clean. The executive director of the Kenmore Square Association, for instance, wrote to Casazza requesting that the city install litter baskets on Commonwealth Avenue in order to enlist "the public, which is not litter conscious," in the project of improving the appearance of Kenmore Square. Casazza quickly installed ten new litter baskets in the square.[24] The Beacon Hill Civic Association made a similar request, as did the Charles River Park Tennis Club in Cleveland Circle.[25] Casazza responded to these requests himself, always in the affirmative. Clean streets were Casazza's job.

In a 1983 profile in the *Boston Globe* significantly titled "In Pursuit of Cleanliness," Casazza admitted that he was "ashamed at the level of [street cleaning] services" he'd "been allowed to provide" during Mayor White's administration. He viewed municipal waste and littered streets as part of the same problem. The key to solving both, he believed, was to increase local waste disposal capacity. In service of this goal, Casazza dreamed of building a local incinerator—which he referred to as a resource recovery plant—that could handle the bulk of the city's municipal waste. An in-city plant would provide Boston with more certainty and more capacity and would free resources for basic city hygiene.[26] In the summer of 1983, Mayor White finally acted on Casazza's advice and solicited proposals for a municipal incinerator.[27]

As the proposals rolled in, Mayor White's administration was drawing to a close. He pushed the city council to sign an agreement with American REF-Fuel, a partnership between Air Products and Chemicals of Allentown, Pennsylvania, and Browning-Ferris Industries of Houston,[28] for the construction of a new waste-to-energy facility at the South Bay location of the city's decommissioned incinerator. Some city councillors had concerns about incineration, but the council overall was attracted to the potentially substantial cost savings.[29] In general, political and technocratic leaders in the city and the state viewed waste-to-energy incineration, which they generally referred to as "resource recovery" (following Commissioner Casazza in the nod to boosters of the new technology), as environmentally superior to landfilling.[30]

Mayor White's successor, Raymond Flynn, took office on January 3, 1984. He was not eager to accept the incinerator solution proposed by his predecessor, citing public health concerns.[31] But as the year wore on, the waste management situation in the state grew more urgent. On July 10, 1984, a parade of garbage trucks from private haulers all over the state rumbled around Boston Common to protest the lack of landfill capacity, which was making it increasingly difficult for them to do their jobs. The state, the private haulers argued, should make it easier to open new land-fills.[32] The protest got the attention of the Dukakis administration, which was already considering a variety of waste issues. The next day, the admin-istration announced a plan that would limit the ability of municipalities to use home rule to reject proposed disposal facilities.[33]

But progress was slow. By 1985, it was clear that there was no consensus among state lawmakers about how to proceed. The public perceived that the state was dragging its feet in providing resources for remediating leaching landfills and for supporting programs or policies that would ameliorate the statewide disposal crunch. An editorial in the *Boston Globe* argued that there had been "too much behind-the-scenes consensus-seeking, and too little leadership."[34] Furthermore, there was some suspicion that the private sector was exercising undue influence in crafting the state policy direction on waste management.[35] Under the guidance of the Department of Environmental Quality Engineering (DEQE), the state was contemplating a "25–50–25 by 1990" plan, under which 25 percent of waste would be recycled, 50 percent incinerated for energy, and 25 percent landfilled. But it took no regulatory steps toward this goal.[36]

In this context of crisis and uncertainty, Mayor Flynn's administration remained undecided about whether to move forward with Mayor White's incinerator plan or to renegotiate the contracts for exporting the city's waste. In early 1985, the state warned the city that the incinerator bid would not meet state air quality standards,[37] but Flynn continued to consider it a viable possibility. Whatever problems the incinerator might pose, exporting waste was considerably more expensive.[38]

Throughout the early 1980s, public officials and the media framed the city's problem consistently: a lack of disposal capacity was driving up waste management costs. In national discourse, popular environmentalist writers were forging links between waste and consumption, but following tradi-tional Sanitary City values, Boston's policy leaders mainly emphasized cost and disposal. Even in conversations about recycling, officials dependably emphasized the stream of material that would remain after recycling. Mayor Flynn frequently reminded reporters that the real work of the city was to resolve the disposal question. In December 1987 he stressed that recycling would not solve the problem:

Flynn asserted yesterday that even if the [state's proposed] recycling center could meet the Dukakis administration's goal of handling one-third of
the city's daily trash output of 1,500 tons, the city would still be left with the
costs of shipping the overflow to existing incinerators in other communities.[39]

A year later, he repeated the same point:

Flynn, in his statement, said that as much as 85 percent of Boston's trash
"will not be recycled under any plan. Boston needs a solution for all of its
trash. We cannot accept recycled concepts. We need real plans and real solutions that address the entire trash issue."[40]

State officials promoted a similar understanding of the city's garbage
crisis. Although some in the DEQE were pushing hard for recycling, the
governor's administration viewed recycling as, at most, a minor compan-
ion to disposal. Alden Raine, a top aide to Dukakis, and Jamie Miller, a recycling expert at the DEQE, consistently emphasized the nonrecycled portion of the waste stream. Characteristic statements to the local press looked
like this one in the *Boston Globe* from August 1987:

Alden Raine, a top Dukakis aide, said last night that the state is building
five recycling centers. But for now, he said, recycling will be able to eliminate
a maximum of 30 percent of the garbage. The rest should go to safe landfills
or be burned in incinerators equipped with scrubbers, he said.[41]

State officials cited a wide range of potential recycling percentages, but
even when citing more ambitious targets, officials emphasized the disposal
portion of the waste stream. For example, a year earlier, the same paper
reported:

Even with an ambitious program ... at most 50 percent of wastes can be
recycled. The rest must be disposed in other ways — preferably in trash-to-
energy plants, Miller said.[42]

At the time, there was not a substantial recycling industry in Massachusetts to advocate for recycling. Voices from the solid waste industry, unsurprisingly, reinforced the problem frames offered by public officials. In particular, representatives of the firm seeking to build the resource recovery
plant repeatedly gave "expert" testimony to local papers about the risks of
pursuing recycling without a reliable plan for resolving disposal. In one of

the more emphatic statements, Peter Watson, a representative of American REF-Fuel, told the *Boston Globe* that aggressive recycling programs had "never been accomplished. It is absolute folly to pretend you can rely on untried means of waste disposal in the face of the enormous quantities that beg for environmentally sound disposal."[43]

The result of this emphasis from officials at all levels was a clear consensus, at least in the public discourse shaped by the local press. "The trash dilemma is straightforward," the *Boston Globe* reported in 1988. "Even if the state someday reaches its goal of recycling 25 percent of its wastes, the rest still must be disposed of. And there are only two ways to do that: Bury it in a landfill or burn it in an incinerator."[44]

The Flynn administration worked actively to frame the solid waste problem in just those terms. Neil Sullivan, Mayor Flynn's policy chief, recalled in an interview that "this was a question of landfills versus incinerators, and this was a question of in town versus out of town." Sullivan understood and promoted the waste issue as a set of two questions: "Should we be responsible for our own trash, or should we send it elsewhere? That's one of the questions. And should we bury it or should we burn it was the other. We tried to frame the community conversation that way. In town or out of town, and landfill versus incinerator."[45]

Mayor Flynn's administration made its key decisions within the framework described by Sullivan. At the beginning of his first term, after some initial hesitation, the mayor decided to pursue a plan of waste export rather than building a garbage burner in the middle of the city.[46] Shortly after that, however, the Flynn administration reversed itself and restarted plans and negotiations for the incinerator. Ultimately, the incinerator proposal could not overcome a variety of obstacles, and at the end of Flynn's first term, the incinerator proposal died again. The administration signed new contracts for garbage export and eventually initiated a minimal curbside recycling program.

At each juncture, the question of garbage in Boston turned on disposal. Approaches that reduced the disposal burden were welcome in discussions, but for decision-makers, solving the disposal problem was their primary responsibility. To be clear, Boston was not at all unique in this regard. All across the country waste managers and local papers framed disposal shortages in similar terms, which emerged straight from Sanitary City logic. Within the Sanitary City regime, garbage was supposed to go "away." As local landfills closed, cities lost these "away places,"[47] which forced local officials to scramble for new ones or risk undermining the whole regime.

Across the country city managers hurried to find new technological fixes or more distant landfills to resolve local disposal crises while preserving the underlying spatial and economic relations of the waste regime. The instability of this moment eventually forced the Sanitary City regime to transform into the weak recycling waste regime. But it is a testament to the power of the Sanitary City waste regime, and the force it exerted in shaping the subsequent WRWR, that municipal officials in Boston emphasized their civic obligation to ensure efficient garbage removal. Policy-makers, particularly at the city level, felt that any alternative waste management strategies, recycling in particular, would be an irresponsible distraction from their primary civic obligation to dispose of waste. As the decision process unfolded, the disposal problem frame was pointedly reinforced by the city-organized waste planning processes.

Maintaining the Disposal Problem: Planning with Limited Expertise

A variety of government agencies and nongovernmental organizations, some with conflicting missions, conducted parallel planning for and analysis of Boston garbage over the 1980s. The processes run by the mayor were dominated by a limited set of interests and expertise. City officials systematically excluded alternative perspectives represented in some of the parallel processes. The result was that the disposal-problem frame remained intact throughout the incinerator negotiations and continued to shape programming even after resource recovery was finally off the table.

The tone for the processes surrounding the resource recovery facility was initially set by the first major planning document, the environmental impact report (EIR). Casazza commissioned the EIR as a narrow technical exercise in the last year of Mayor White's administration, before American REF-Fuel had been formally contracted. The report was drafted by CSI Resource Systems, a nationwide incineration consultant that, not coincidentally, was also advocating for incineration in Seattle at the time. Although CSI's EIR fully evaluated only one option—a resource recovery facility at South Bay—the final section of the report briefly considered four alternatives: hauling Boston's waste to landfills outside the city; hauling Boston's waste to incinerators outside the city; building a landfill in Boston; and building a WTE facility elsewhere in Boston. Each of these options received only enough attention in the report to demonstrate why it would be less desirable than the South Bay plant. CSI described each option as "uncertain" and "costly." The firm argued further that "the practice of long haul to distant landfills has associated environmental impacts that will be significantly reduced if the proposed Facility were to substitute for this practice." The environmental impacts that would be reduced, according to CSI, were related to truck traffic and groundwater contamination. The EIR presented the South Bay facility as a "long-term economically and environmentally sound solution to its solid waste disposal problem."[48]

There were some interesting omissions in the EIR. For instance, the document did not mention ash disposal. It also contained no documentation of public comment and no discussion at all of recycling, reuse, or waste reduction as possible companions to, or replacement for, incinerator capacity.

The EIR did, however, include an artist's rendering of the proposed facility. The rendering was purely schematic; the facility had not yet been designed. Thus, its presentation in the report can be understood as a conceptual signal about what the facility would represent more than what it would actually look like. In the rendering, the plant is presented as an innocuous, even boring, office building on a fully landscaped site. Off to the side of the facility, there is a single white smokestack projecting up into a pink sunset. It looks more like the Bunker Hill Monument than an incinerator stack. Behind the facility, just visible, is the top of the recently constructed Prudential Building, an important symbol of progress and development in Boston. The angle of the representation makes the facility look like a guardian of the modern, growing city behind it. The WTE facility looks like a thoroughly unobjectionable contribution to the city (figure 2.1), especially when compared to the preexisting incinerator (figure 2.2).

The Flynn administration moved into City Hall just as the EIR document was released. Flynn and his staff prided themselves on transparent decision-making. The press had access to Flynn's staff and the mayor himself as well as to many waste-related meetings and events over the course of the decade. The administration even established its own formal process for making a decision about the South Bay incinerator. The administration's process, though transparent, was narrow. It built directly on the initial option laid out in the EIR.

A key element of Flynn's planning process was the appointment of a Citizen Advisory Committee on Solid Waste Disposal. Flynn appointed the committee shortly after his inauguration, and it remained active during the period when the administration was making its first decision about whether to move forward with the South Bay resource recovery facility. The limited nature of the committee's role, however, is evident in its title, which emphasizes disposal rather than a more inclusive concept of waste management.

The advisory committee included eleven full members and two ex officio

Figure 2.1. Artist's rendering of the proposed South Bay facility, included in the 1983 environmental impact report (EIR); no other option received serious evaluation in the document.

members, including Jamie Miller from the Massachusetts Bureau of Solid Waste Disposal and Robert Travaglini, a Boston city councillor who was then chair of the Committee on Urban Resources. The full membership included representatives from several Boston neighborhoods closest to the incinerator site, including the South End, Dorchester, Roxbury, Fenway, and South Boston. The advisory committee also had three members representing Boston's business community: Samuel Tyler, president of the Boston Municipal Research Bureau (BMRB, a local economic development think tank); Nicholas Georgenes, president of the New Market Business Association; and Richard Mori, director of metropolitan affairs for the Greater Boston Chamber of Commerce. Betsy Johnson, a representative of the South End neighborhood and also of the American Lung Association, was the only member with a stated connection to a public health organization. None of the members had recognized affiliations to any environmental or civil rights organizations.[19]

The advisory committee was given a narrow mandate. The mayor instructed its members to comment on only three options, which seem to have been drawn from the EIR: a proposal to haul waste out of Boston; the American REF-Fuel WTE proposal; and a third proposal to haul all of Boston's waste to a private landfill in Plainville, Massachusetts. It was a further limitation that the committee was advisory only. Members were instructed to provide comments on the specified options, but to make no recommen-

Figure 2.2. South Bay incinerator, Boston, 1970, by Spencer Grant. Courtesy of Spencer Grant.

dations. The committee would have no formal role in the administration's decision and no purview to recommend or review alternatives beyond the three assigned.[50]

The committee's charge asked only for an assessment of issues related to the three options. Committee members evaluated each of the three proposals according to preestablished criteria and identified a host of issues that merited further consideration. Traffic and cost, two issues of particular concern to the city's business community, received the most attention in the report. Public health also received a good deal of attention, reflecting the interests of neighborhood representatives. Though compatibility with recycling was a permitted evaluation criterion, in its assessments the committee did not develop the issue of recycling in relation to the proposals. In terms of environmental concerns, the report noted explicitly that the committee lacked sufficient data to make substantive comparisons. The committee members recognized that any environmental impacts of incineration would apply whether a facility was constructed inside or outside the city of Boston. This point indicated a concern for the region that was not present elsewhere in public debate.

Committee members felt hamstrung by the limited mandate. In a signed introduction to the report, they expressed concern about the lack of public participation in the evaluation of the options and frustration that they were not able to explore other alternatives—even minor ones, such as the possibility of a smaller Boston-based facility.[51] After the president of the Massachusetts Senate, William Bulger, in an overtly political move, publicly opposed the incinerator in 1987 (more on this in a moment), exasperated committee member Betsy Johnson vented to the *Boston Globe*: "If this is what it took for the city to finally sit up and take notice of what we've been saying for three years, I say more power to President Bulger."[52]

The advisory committee was the only mayor-sanctioned group publicly evaluating waste management options for the city. But many other government agencies at the city, regional, and state levels were also studying the city's garbage. Several documents assessing Boston's municipal waste situation were released between 1984 and 1987. Though clearly these were relevant to the city's decision, the mayor did not consider them.

During the local elections of 1985, two city council hopefuls campaigned on the incinerator issue and got elected. They had argued to a receptive public that there were better options for waste management than incineration.[53] These councillors, Rosaria Salerno and David Scondras, pushed the city council to propose an alternative vision for Boston's waste system. They brought in recycling and environmental advocates for public hearings and committee meetings about Boston's options. As part of their efforts, they also brought in speakers to educate the council on options beyond disposal. Barry Commoner, a nationally known incinerator opponent and recycling advocate, testified before the city council's Special Committee on Solid Waste, a subcommittee of the council's Environment Committee, in December 1988. He presented information on recycling, arguing that the city should be ambitious. Based on his assessment of waste in other places, he argued that 90 percent of the waste stream was recyclable in some fashion. Commoner identified key conflicts between recycling and incineration and argued that recycling could be set up more quickly and cheaply than an incinerator with good results. He also argued for rational planning, evaluation of the options, and selection of the most sensible approach. At the same hearing, Stephanie Pollack, from the Conservation Law Foundation, also testified, and she, too, advocated for recycling rather than incineration as an alternative to landfills.[54] Both Commoner and Pollack positioned incineration as a dangerous and unproven technology that would discourage recycling and impact the city for years to come.

A week after Commoner and Pollack made their presentations, on December 8, 1988, the city council's Special Committee on Solid Waste unveiled its own plan, called "Dealing with Our Trash." The report, authored by Scondras and Solerno with the help of Harvard's Kennedy School of Government, included an assessment of Boston's current waste management predicament and its options. Scondras and Solerno proposed a waste management program for Boston that channeled the EPA's best intentions for ISWM. The plan proposed a completely new solid waste system for the city that would have included legislation to reduce the use of toxic chemicals in consumer products, mandatory recycling and composting, and a pay-as-you-throw (PAYT) rate structure to encourage consumption behavior change. Resource recovery through incineration, they suggested, should be a last resort reserved only for unrecyclable material.

The summary diagram for the plan illustrated the dramatic departure of the plan from the status quo. The first diagram, in an unmistakable critique of a wasteful and unsustainable one-way flow, pictured the current material flow through the economy (figure 2.3). Inputs come from elsewhere, are consumed, and are then thrown away. This was how the WRWR was organized. The councillors' plan was an ambitious challenge to that ordering. The second diagram illustrated several loops, with materials being captured through various processes and reintroduced into the economy (figure 2.4). If Boston had adopted the plan, it would have mounted a challenge to the production of disposability.

CITY OF BOSTON
CURRENT WASTE DISPOSAL SYSTEM

Figures 2.3 (above) and 2.4 (opposite). "Dealing with Our Trash: A Report to the Boston City Council," Boston Special Committee on Solid Waste Management, October 3, 1988. Figure 2.3 depicts Boston's current waste system, and figure 2.4 shows a potential ISWM system.

The plan was not abstract. The councillors considered the institutional requirements for implementing such a radical shift in waste policy. The plan included a draft city ordinance that emphasized recycling, composting, and waste reduction, with disposal in any form only for nonreusable, nonrecyclable materials. The plan also proposed to establish a solid waste commission to be headed by the public works commissioner—with representatives from neighborhoods, industry, and environmental groups—that would oversee an extensive waste planning process and monitor progress toward solid waste goals. The commission would levee fees, in what was essentially a PAYT scheme, to ensure a revenue stream to pay for recycling programs, and it would develop a system of fines to ensure compliance.[55] That revenue would then fund a solid waste authority with more flexibility to innovate and develop new programs.

The report's authors acknowledged contributions from many local stakeholders, including the mayor; his policy chief, Neil Sullivan; and Public Works Commissioner Casazza. But when the report went public, the Flynn administration publicly demurred. The plan was likely politically useful for Councillors Scondras and Salerno, but it went nowhere in terms of actual policy implementation. It conflicted with Casazza's vision and sidestepped the administration's in/out, burn/bury problem framing. It would have required a lot of buy-in to even start the conversation.

Both Casazza and Flynn's aides dismissed the report out of hand. The *Boston Globe* reported that neither Casazza nor Flynn's environmental lead, Robert Bauman, had been consulted in the development of the plan. Bau-

INTEGRATED WASTE DISPOSAL SYSTEM

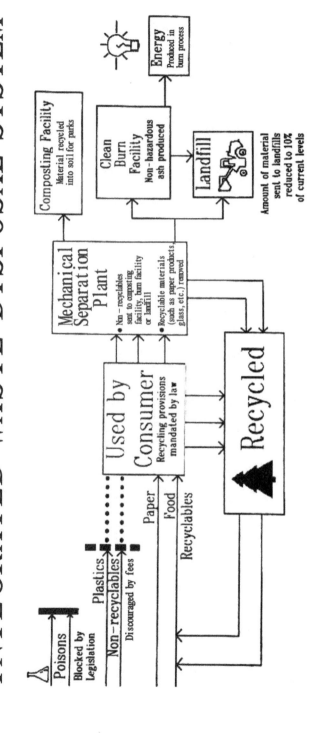

man told the press that the plan had "serious flaws" and presented an "enforcement nightmare." Casazza refused to comment, telling reporters he hadn't even had time to read it.[56] Mayor Flynn also publicly criticized the plan, arguing against the whole vision, but with particular venom for the fee-for-service model. Councillor Scondras argued that the fee and fines system was just a set of tools that would "give the administration all the power it needs to start a real recycling program." But Flynn told the papers that the system was "needlessly complex and controversial."[57] The city council approved the mayor's budgets, but otherwise had no direct role in solid waste policy-making. The plan had no power. Flynn and his deputies hastily dismissed it and moved on.[58]

Other government agencies were also planning for solid waste. The Metropolitan Area Planning Council (MAPC) released a regional plan for Metro Boston's waste in 1986. This document contained contextual information about solid waste in the region, including a detailed town-by-town assessment of waste generation, collection, recycling, and disposal. The MAPC concluded that the cost of landfilling would likely continue to increase, and thus towns that were able to do so would be wise to invest in resource recovery. The MAPC also recommended that each city and town initiate recycling and source reduction programs "to cut costs and conserve landfill space."[59] The report, which contained detailed profiles of MAPC-region cities and towns, showed that nine of Boston's neighbors already had recycling programs, indicating that some recycling expertise and infrastructure already existed in the region. It also contained a detailed waste composition analysis for the region.

Unlike the Boston City Council plan, the MAPC's recommendations were framed in WRWR logic. The report did not deviate from the assumption that the primary function of recycling was to reduce the disposal burden. This was the narrative of the disposal industry side of the waste regime. Though that's the framing that would likely have resonated with Boston decision-makers, the report is not reflected in any extant city documents.

At the state level, the DEQE had a solid waste division that was actively working to design a progressive state agenda for recycling infrastructure and to establish an official position on incineration, and other state offices were also weighing in. In the early 1980s, prior to the peak of interest in waste planning at the city level, the state's Science Resource Office released a report called "Solid Waste Management in Massachusetts," and Governor Dukakis released a waste management plan for Massachusetts in 1985. These planning processes were not, however, coordinated with the City of

Boston. None of the city's planning documents contain any references to the work happening at the state level or by the MAPC.

Boston's business community was actively analyzing the city's solid waste situation as well; indeed, of all possible stakeholders, the business community was the group best represented in the city's decision-making, given the inclusion of three key business leaders on Flynn's Citizen Advisory Committee on Solid Waste Disposal. In addition to having an institutional voice on the advisory committee, the Boston Municipal Research Bureau, a think tank primarily devoted to serving Boston's business community, worked outside of institutional channels to shape opinion and influence decisions. BMRB representatives were frequently quoted in the press. As representatives of private business, BMRB analysts focused on costs, efficiencies, and the impacts of the city's decisions on the commercial sector. The BMRB was generally supportive of the privatization of waste collection and treatment. It argued that the city should consolidate public contracts and provide increased waste disposal capacity to relieve the increasingly tight capacity in the region.

In 1982, before Mayor White had released the request for proposals for a South Bay WTE incinerator, the BMRB published a report arguing that investing in resource recovery would allow Boston to avoid becoming "a captive of the market."[60] This message ended up being a key part of Mayor White's initial justification for the South Bay proposal. By virtue of the cost differential, particularly for the commercial sector, as well as the opportunity for the city to exercise more control over disposal and the regional disposal crisis, the BMRB strongly advocated for the American REF-Fuel proposal. Short of that, the bureau recommended that the city facilitate the consolidation of all municipal waste collection so that the commercial sector would not remain subject to the "open market," which could result in substantial cost increases over the coming decades.[61]

When Mayor Flynn initially decided against White's incinerator proposal in 1986, instead opting to export the city's waste via a twenty-year contract with the hauling firm Signal Environmental Systems, BMRB released a report arguing that in-city resource recovery would cost the public sector over $90 million *less* than the export plan and save the private sector $120 million in trash hauling and disposal contracts.[62] The report made a splash in the press, and the *Boston Globe* ran articles insinuating that waste export was the wrong decision.[63]

Flynn quickly reconsidered the matter and reopened incinerator negotiations. Over the 1980s, the BMRB strongly and influentially advocated for WRWR solutions. The organization prioritized efficiency and cost, and never questioned the concept of increasing disposal capacity as the obvious solution to a disposal problem. And apparently, it had the ear of the mayor.

Despite the range of views expressed in the many plans and analyses published during the mid-1980s, only a limited set of influences—including Casazza and the BMRB—actually shaped how city decision-makers viewed their options. The dominance of these insider voices reinforced the city's disposal-problem frame and kept the debate narrowly focused on whether disposal would occur inside the city or outside of it, and by incinerator or landfill.

City officials' narrow approach to framing the problem and its preoccupation with business-sector perspectives should not be taken to mean that there was no public interest in waste. The proliferation of plans and analyses documented above demonstrates a high level of public interest in solid waste management and planning in government and beyond. The city council's ISWM plan was broadly responsive to issues raised during the city council campaign, and at various points public hearings held by the mayor's office and the city council were flooded with South Boston residents and neighborhood leaders who opposed the incinerator. When Flynn formally announced his intention to proceed with the incinerator in May 1986, for instance, community leaders showed up in his office to express frustration about the mayor's indecisiveness, the potential health impacts of the incinerator, the noise and truck traffic it would bring to the area, and the fact that the site could be more profitably developed with a different industrial use.[64] Betsy Johnson, the resident of nearby South End and member of the Citizen Advisory Committee, frequently attended public meetings and complained to the press about the lack of a comprehensive process. "Did the city ever ask itself," she once probed, "if it could have gotten away with a smaller incinerator, or one in another area?"[65]

In addition to opposition from neighbors, advocacy groups like Fair Share Massachusetts and Greenpeace sustained campaigns to push the Flynn administration to abandon incineration and develop an aggressive recycling program. Greenpeace activists regularly attended city meetings, arguing that the incinerator was basically "a toxic ash factory," and making sure it was publicly known that Greenpeace intended "to fight [it] to the wire."[66] Neil Sullivan, Flynn's lead on solid waste, had actually been a staffer at Mass Fair Share prior to joining the administration, and the press regularly reported on meetings between the administration and advocacy groups. Within formal planning documents, though, there is little evidence of citizen or advocate voices. As residents and activist organizations raised the alarm, Boston officials responded that "the city has a trash problem it has to deal with now, immediately, and this is our best remedy."[67]

Ultimately, Flynn's administration made decisions with sincere intentions, but was limited by its own narrow way of framing the problem. The voices and the knowledge that the administration allowed into its formal decision-making process reinforced Casazza's Sanitary City values and created a strong foundation for implementing WRWR waste management strategies. Recycling was an afterthought; the city's responsibility was disposal. A second feature of Boston's planning served this same purpose: throughout the decision-making process, there was very little, if any, engagement with waste itself.

Garbage as a "Relational Object"

Throughout the 1980s, garbage remained a "negative/relational" object, in geographer Sarah Moore's terms.[68] The city's waste was not defined by any quality inherent to itself; instead, its meaning derived from its constitutive relationships with other objects and priorities. Although waste managers and contractors likely did have some knowledge about waste composition and generation, those material details did not play a role in waste management decisions. The EIR, for example, does not contain any documentation about waste composition or expected generation trends. This is a surprising omission, because the efficiency of resource recovery operations can depend on the specific material mix within the waste stream. Recycling advocates would also point out that a realistic assessment of the waste stream would have revealed a much higher potential for recycling than the city ever acknowledged. In the absence of a direct assessment of garbage itself, economic development, a court-ordered clean-up of Boston Harbor, and political rivalries informed key decisions.

The politics of jobs was a key theme throughout the planning process. Mayor Flynn made economic development and jobs for residents "the highest priority in his administration."[69] Any industrial development of the South Bay site was, not surprisingly, evaluated in these terms. At each turn, Flynn justified his administration's current thinking about resource recovery by framing it in terms of the impacts to business, industrial development, and job creation.

For example, after Flynn made his first decision, in August 1985, to abandon the incinerator plan and export waste, he was asked by reporters how he justified the more expensive proposal, given the city's budget crunch. The *Boston Globe* reported Flynn's answer:

> I probably arrived at the decision the other morning at breakfast with many members of the business community who were on hand. And we discussed the finances of both of these options.
>
> And it was never my contention that cost alone was going to be the overriding ... consideration for the siting of the waste-to-energy facility.
>
> Many people concerned about the economics of Boston indicated to me it would be very difficult and might be shortsighted to just consider the cost of the per tonnage rather than looking at the long-term negative consequences that a more proper utilization of that land (in the South Bay) may have. You're talking about 7.5 acres of prime potentially exciting industrial property and you're talking about a grand total of 50 permanent jobs (in the in-city plant).
>
> We're trying to expand our economy so people in those areas have jobs. And I thought, based on that kind of enlightened look at it, along with these other factors, such as the environmental and neighborhood impacts and so forth, that the cost factor alone would be shortsighted.[70]

In this statement, Flynn indicated his concern about long-term economic development and reinforced the key role of business interests in decision-making. Environmental concerns, neighborhood impacts, cost, and the garbage itself were all ancillary to job creation and economic development.

Flynn was not alone in this ordering of priorities. His Citizen Advisory Committee reinforced the perspective. The committee's report argued that "construction of an in-town plant would create temporary construction jobs. Operating the facility would create 42 permanent jobs. However other development of the site could create more jobs."[71] Residents and community leaders also felt that a different kind of industrial development could likely yield more jobs and more revenue to the city than an incinerator.[72]

The administration's concern for jobs development was formalized in the city's negotiations with American REF-Fuel, while Flynn was still considering the South Bay plant. Part of the REF-Fuel agreement was a commitment to conform to Boston's resident job policy, which prioritized Boston residents for public jobs.[73]

Economic development and job creation are typical drivers of local politics. It is significant here not because it represents any departure from normal urban priorities, but because of how the concern for jobs and economic development supplanted other relevant considerations that might have accompanied the decision to invest in such expensive public infrastructure. City decision-makers defined garbage in relation to its ability to create jobs. The emphasis on economic development in the Boston case reveals how garbage was viewed as a means to an end rather than something in and of itself. As other infrastructural priorities crystallized for the Flynn administration, they reinforced this sense of immateriality. The real death knell for the incinerator came finally when pressure to make other progressive infrastructural investments literally displaced the incinerator from its prime location in South Bay.

In the 1980s, Boston Harbor was notoriously polluted. Before Boston constructed its first primary treatment apparatus in the 1950s, the harbor had been receiving a daily discharge of raw sewage for eighty years, ever since the city's first sewerage collection system had been completed in 1876. Even after the 1950s, the city's sewage was only minimally treated, and rainstorms frequently overwhelmed the existing facilities, resulting in overflows of untreated sewage. A century of raw sewage, plus decades of unregulated industrial pollution, turned Boston Harbor into one of the most polluted bodies of water in the United States.[74]

Boston managed to win exemptions from increasingly stringent federal water pollution standards throughout the 1970s, allowing the harbor to remain fetid well after other cities had initiated cleanup procedures of similar bodies of water. A series of lawsuits initiated by the City of Quincy and the Conservation Law Foundation in the 1980s finally culminated in a court order to clean the harbor. In an unusual ruling, Massachusetts Superior Court dictated the creation of the Massachusetts Water Resources Authority (MWRA), a regional body charged with the management of water and sewage systems. The primary function of the MWRA, initially, was to construct a new primary and secondary treatment system that could handle the growing volume of sewage in the Boston region and nurse Boston Harbor back to health.[75]

The MWRA decided to replace the inadequate sewage facility on Deer Island with a high-capacity plant that could serve the entire region. Other options, such as a plant in a different location, or a network of smaller plants, were deemed infeasible for political, fiscal, and geographic reasons.[76] The Flynn administration was supportive of the efforts to clean Boston Harbor, and it was a top priority of Michael Dukakis's gubernatorial administration as well.[77]

While the solution was optimal for many reasons, there was one key obstacle: Deer Island was full. Since the mid-nineteenth century, Deer Island had been Boston's preferred location for undesirable public functions. It had housed many institutions, including a poorhouse and a quarantine station. By the 1980s, it was home to the Suffolk County House of Correction, a century-old building that had never been designed for long-term residents. The correctional facility had become an overcrowded prison in which people were subjected to dreadful conditions for the duration of sometimes lengthy sentences. Boston managed the county facility, and improving conditions for inmates was another key priority of the progressive Flynn administration (see figure 2.5).[78]

To make things more complicated, although Deer Island is technically within Boston's jurisdiction, land access to the peninsula is only available through neighboring Winthrop. Winthrop, which also endures some of the greatest impacts from Logan Airport, was resistant to hosting access to a massive new sewage facility as well as an upgraded and enlarged prison. As part of a bargain with Winthrop's city council, the Commonwealth of Massachusetts, the MWRA, and the City of Boston agreed to relocate the house of correction.[79]

A new site for the facility was quickly identified: South Bay. From the point of view of politicians, the same characteristics that made South Bay a perfect home for an incinerator also made it an ideal site for a prison. It had no residential neighbors, it was owned by the city, and it was centrally located and accessible to transportation infrastructure. The Flynn administration was eager to improve the inhumane conditions at the facility and liked that the location would give easier access to families, service providers, and lawyers.[80]

Given the urgency of the harbor cleanup, state leadership eagerly supported the prison relocation plan. The Dukakis administration offered the city $40 million to help defray the costs of the relocation. Initially, Boston officials believed that both the prison facility and the incinerator could fit on the site. Despite the administration's genuine distress about living conditions in the outdated facility, they saw no conflict—other than space constraints—about co-locating the public facilities. Eventually, it was determined that shoehorning both facilities into the area would be impossible.[81] In the end, the key product from the infrastructural negotiation was another relational identity for garbage, this time as a public nuisance infrastructurally equivalent to sewage and incarcerated people.

Garbage was defined by its ability to generate jobs. It was defined by its pragmatic equivalency to the city's incarcerated population. As incinerator negotiations unfolded, garbage also became a new public arena for a longstanding feud between Mayor Flynn and Massachusetts Senate president

Figure 2.5. A = The South Bay site; B = Deer Island. Base Map: Google, TerraMetrics.

William Bulger. Given that Bulger was from South Boston, his opposition to the South Bay facility might have been expected; it is not uncommon for politicians to oppose the siting of waste infrastructure in their districts.[82] But city-state political relations, particularly as they played out between Flynn and Bulger, were not just about infrastructure siting. More abstractly, garbage became a medium through which political power was tested and expressed.

During the first few years of Flynn's administration, Bulger had remained quiet on the disposal crisis and on the incinerator. But on July 16, 1987—after Flynn had already been working on the trash issue for nearly two years—the Senate president stunned the Flynn administration by announcing his opposition to the incinerator publicly, without warning, and then filing a surprise bill in the state legislature to prevent the city from siting the incinerator at the South Bay location.[83] The following day, Flynn took the unusual move of calling a press conference on the State House steps to protest Bulger's bill. Bulger showed up to the conference, and the two politicians, both known for sedate public performances, went toe-to-toe in full view of the Boston press corps. Flynn defended the incinerator and charged Bulger with dirty politics. But Bulger stole the show by announcing that he had a "near perfect" site in mind for the incinerator—outside the city—but would not disclose where.[84]

Bulger underscored the political nature of his objection to the South Bay proposal by eventually offering wildly unrealistic alternatives. After months of mystery, he finally announced his proposals, which included technologies—such as an incineration-at-sea—that were widely discredited. His primary alternative was to send Boston's trash to a recently closed quarry in the wealthy suburb of Weston.[85] Aside from the fact that Boston officials had no jurisdiction over the Weston site, the proposal was politically preposterous. Within months of Bulger's announcement, Weston rezoned the site out of industrial use.[86]

To block the incinerator, Bulger used all the tools at his disposal. In addition to hijacking headlines, he used his control over the Massachusetts Senate to hamstring, and eventually eliminate, the $40 million grant promised by the Dukakis administration to offset the costs of new infrastructure related to the prison relocation and for industrial development.[87]

After Bulger announced his opposition, he and Flynn did not communicate directly about the incinerator or alternatives. The historical record suggests that Bulger and Flynn met face to face on the issue of waste management only once—and it was during the "storm on the State House steps."[88] The lack of substantive engagement between the two politicians suggests that neither was focused on the task of solving the waste crisis. Bulger's opposition may have been grounded in legitimate concerns about potential impacts to his South Boston district, but it was understood by the Flynn administration and the media to be more about political rivalry than substantive objections to resource recovery in Boston.[89] Both from South Boston, Flynn and Bulger were well known for their competing political ambitions and lifelong antipathy toward each other. As one staffer from the Flynn administration put it:

> Flynn and Billy Bulger had been knocking heads going back to the early '70s—because Ray Flynn did not play by the South Boston political rules and Billy Bulger was in charge of the South Boston political rules. So now that they both had significant power, at various times they did their best to thwart each other. I'm one side of that. I'm trying to say it so that it's down the middle, but there it is.[90]

The Flynn-Bulger feud was good fodder for local news, and each politician exploited the media-ready stand-off to achieve his own ends. Bulger garnered headlines for savvy tactics and wild proposals all in the name of protecting his district. It is likely that Bulger also liked the idea of creating a high-profile failure for a longtime political rival. Flynn, for his part, was able to lay blame for Boston's challenges on inconsistencies and lack of support from the state government.[91] Antics aside, the centrality of the political story demonstrates how garbage became a medium for testing and expressing political power.[92]

While the Bulger-Flynn feud played out in personal and political terms, Flynn also used the incinerator to highlight what he viewed as the failure of the state government to provide solid waste solutions to the city. Mayor Flynn wanted the state to provide a stronger framework and more resources. After it became clear that the incinerator would not go forward, Flynn made a point of arguing that the state should be playing a more proactive role in developing disposal capacity and planning for waste management at the state level. As the city considered a state proposal for a recycling facility to accompany the Suffolk County House of Correction at South Bay, Flynn told the *Boston Globe* that the facility could yield a nationally significant recycling program, but that this was "no panacea for the trash problems we're going to face in the future." "Hopefully," he added, "the state will come up with a comprehensive solid waste program."[93]

Within the narrow in/out, burn/bury problem frame, the potential power in garbage lay in who got to decide among the prescribed options. Bulger technically won on that count even as Flynn characterized the decision as a failure of state government. And, given the narrowness of the debate, the victory was also narrow. As the politics unfolded, all of the city's garbage hung in the balance. And yet no one making decisions even really knew what that garbage was made of. Throughout the decision process, garbage remained an abstraction, reinforcing the assumptions, definitions, and patterns of the WRWR.

The Immateriality of Garbage and the Stability of the Disposal Problem

Throughout the decade, officials and the press talked about garbage in limited terms. In the press, it was simply trash, waste, or garbage. These words are evocative, but vague. The language of "resource recovery" is vaguer, even misleading. It signals a kind of efficiency sure to appeal to a thrifty Yankee sensibility and suggests treasure from trash, something from nothing. It is the language of engineers who seek efficiencies within the artificial boundaries of the end-of-the-pipe. It is the language of corporate marketing professionals trying to sell a product. The term does not reveal anything about the process that underlies the trash-to-treasure alchemy. It does not tell us that household waste materials—specific items with specific properties that were created and marketed by specific interests—are going to be converted to energy through a relatively inefficient process of burning everything together—and that involves adding other fuel when the mass is too wet or too inert to burn on its own. It does not indicate that the process will necessarily create toxic emissions in solid and gaseous form. And it certainly does not expose the uncomfortable fact that after the incineration process is complete, the city will be responsible for disposing of the often highly toxic ash—still a substantial volume—in a landfill somewhere.

The city did acknowledge the potential air emissions from the resource recovery process in the EIR—though, according to the state, the assessment was woefully inadequate even there. But the EIR did not consider ash disposal. Public concerns about incinerator ash provide a peek into how waste and its outputs were abstracted and dematerialized through the planning process. As noted previously, many constituents were concerned about ash. Greenpeace activists from Boston tried to convince the EPA to declare incinerator ash toxic waste. One of the activists' actions involved bringing a container of ash from an open ash dump in Saugus—ash which was the product of the incinerator where much of Boston's waste was being treated—to the EPA's regional offices. The activists actually scuffled with EPA security, but they were eventually admitted to the office with a small container of ash in tow.[94] This action engaged the actual substance of incinerator ash, with Greenpeace asking regulators to consider what was in it. The activists were trying to introduce the materiality of garbage into the policy discourse.

But Flynn's administration was not receptive to this dialogue. His advisers recognized abstractly that ash could have health and environmental consequences.[95] A researcher from the Boston University School of Public Health told city public health regulators that ash was highly toxic and therefore "must be disposed of in an environmentally sound manner,"[96] but he provided little indication of what this might mean in practice. (He also argued that because of the toxicity of ash, any incineration should be accompanied by investment in recycling—another argument for recycling that got sidelined.)

But Boston officials, while recognizing the toxic quality of ash in the abstract, did not anchor this to a more grounded or materialized assessment of garbage incineration. For example, to plan for ash management, city officials relied on their private-sector incinerator partner, American REF-Fuel. REF-Fuel's project manager, Peter Watson, assessed a potential ash disposal site in the nearby city of Quincy. It was clearly in the company's interest to identify an affordable ash disposal site. Using the abstract language of engineers, veiled through REF-Fuel's corporate interest, Watson detailed his assessment in a memo to Casazza. First, the memo refers to ash as "residue," abstracting ash completely. Watson described the site as "imperfect" but "relatively suitable" and noted that it offered an "opportunity for curtailing costs."[97] This is the closest the memo came to a physical description. The terminology and the content of the site analysis negated the problematic material properties of ash completely. It also ignored the political and social identities of the ash, as well as the waste material it came from. It framed the project of waste management as a technical exercise grounded in engineered risk management and cost minimization, and it suggested that solutions that maximized profit could be optimized through the use of technology.

Waste itself was treated in similarly conceptual terms. The city's 1983 request for proposals for the WTE facility provided a paragraph on the total tonnage of garbage, with a table of collection tonnage by district and by month for the previous year as well as a map of the city's collection districts.[98] That was it in terms of material or flow analysis. In 1985, public works assessed the potential of recycling in Boston by estimating the amount of potentially recoverable materials in the waste stream based on an eight-year-old waste composition study from the neighboring city of Somerville.[99] Recycling experts from outside the city offered general observations about the percentage of recyclables in municipal waste gathered in other cities.[100] Flynn's advisers had determined, by guesswork, that about half of Boston's waste was commercial and half residential.[101] Frequently, advisers or members of the Flynn administration would cite aggregate numbers about the volume of Boston's waste. These general statistics, based mostly on estimation and inference rather than direct observation or measurement, left waste as an abstract, unknown mass.

The non-examination of garbage represented a kind of "unseeing," in Robin Nagle's terms.[102] Because city decision-makers could not see the materials themselves, they were not able to question definitions and assumptions framed by the waste regime. Therefore, garbage remained defined in WRWR terms: something to be gotten rid of; a substance determined more by the necessity of its removal than by its own inherent qualities or content. By accepting and reproducing this definition, the city reinforced the emergent weak recycling waste regime, giving it substance through municipal policy and practice.

The Stability of the Disposal Problem through Crisis

Dominique Laporte famously quipped, in his seminal *History of Shit*, "Surely, the state is the sewer."[103] By this he meant that through the construction and operation of sewers, the state organizes space and behavior, and thus asserts its power. The play, of course, is that government is itself a murky and foul arrangement. In spite of the Flynn administration's well-meaning progressivism, Boston's politics of garbage displayed both meanings in equal measure. Decisions about waste management in the 1980s were one part power politics and one part biopolitical Sanitary City technocracy. The decisions were hugely consequential for citizens, but the decision-makers avoided citizen engagement.

The city's technocratic and political waste planning allowed waste to be defined relationally. Unlike the sensory and material characteristics of sewage, or the physical conditions of the Suffolk County House of Correction, both of which were well documented and even litigated, garbage in Boston was defined only in its relation to other materials and other priorities. It was understood variously as a nuisance equivalent to sewage or incarcerated people; as a medium for channeling state power; as a potential source of jobs and industrial development; and as a substance to be gotten rid of. It was not relevant to decision-makers that these definitions were somewhat contradictory. Because waste itself had minimal substance, it took the shape of each new political context.

While the definition of waste shifted mercurially, it never materialized in a way that could destabilize the disposal-problem frame. Between 1984, when he took office, and 1988, when he finally agreed to initiate a city recycling program, Mayor Flynn downplayed the potential of recycling in strikingly consistent terms. He always maintained laser focus on disposal, arguing that recycling should not distract from that fundamental responsibility. Thus, for over half a decade, through plans, competing infrastructural demands, and political crises, Boston officials were always trying to solve a disposal problem.

It is important to note that officials working for the Flynn administration, including Mayor Flynn himself, genuinely wanted to reach the right decision. As limited as the process was, members of the administration wanted desperately to act responsibly. After they had made the final decision to export garbage instead of building a local incinerator, Neil Sullivan remembers being *"totally* relieved":

> I didn't want to put an incinerator in South Bay. Every time I drive the 'Pike and go by the Millbury plant, I think, "My god, I can't believe we were going to put a thing like that in South Bay." But I also understood why we ended up there for a while. And that was, we could not control either the price of the trash disposal or the environmental impact of it. Don't underestimate how attracted we were to the latter, taking responsibility for the environmental impact of Boston's trash. It was something, you know, we policy junkies were actually real interested in, even if it was a fool's errand perhaps. But we felt we should be responsible for our own trash.[104]

The members of Flynn's staff believed they had a moral responsibility not to impose the city's garbage on another community, if possible. This conviction permitted them to consider an unpopular, and perhaps unwise, disposal option. Without an alternative view of what trash was — an externality of unregulated production, perhaps, or the product of overconsumption, or a stream of materials that could be used in other ways — only certain limited calculations were possible. The WRWR, channeled through trusted city experts such as Joe Casazza, provided baseline definitions and assumptions to bound the city's options, and the city's narrow process reinforced those boundaries. The whole process produced the newly forming WRWR within the city of Boston. And, as city officials realized, the WRWR offered no great solutions. In the WRWR, there will always be too much garbage, and it always has to go somewhere. Waste reduction was never on the table, and recycling only emerged as a serious possibility after the incinerator question had been settled and the city had already formalized its contracts for waste export.

By shutting citizens and activists out of waste planning, relying on the expertise of waste professionals and businesses, and actively holding to a narrow Sanitary City problem frame, city officials advanced the interests of the WRWR, ensuring that city policies and practice would reinforce and reproduce the waste regime locally. The limited decision process meant that the full range of voices, knowledge, and values evident in the discourse around waste in Boston was never leveraged to generate a full range of alternatives. Boston leaned on preestablished and relational definitions of garbage and then built a system to manage it that effectively realized the weak recycling waste regime.

By 1988, Boston had abandoned the incinerator idea and committed to exporting all of its waste to regional facilities. The mayor signed a recycling ordinance (to be explored in detail in chapter 4), but he made no concrete plans or progress toward waste reduction or actual recycling programming. These decisions put Boston squarely in the mainstream. A few cities did build incinerators, but most of them made determinations similar to Boston's, emphasizing regional export as a solution to local disposal shortages. The path that Boston took to arrive at these decisions shows how WRWR objectives manifest locally.

CHAPTER 3

Deconstructing Garbage: Radical Reframing in Seattle

Seattle is "urban by nature," a bustling city, exploding with tech-industry growth, nestled among magnificent mountains and waters.[1] The area's earliest colonists wrested land and resources from indigenous inhabitants, hoping to capitalize on the region's abundant natural resources. Over time, colonial "engineers with vision and courage" transformed the land, block by block, creek by creek, to protect their newly urbanized territory from frequent flood and fire.[2]

Settler efforts to control, subdue, and, eventually, protect nature through infrastructure remain a visible part of the city's identity. Seattle is a messy and mixed metropolis that exposes the fragility of the boundaries between the urban and the natural in Western thought.[3] Urban development has destroyed many of the fragile ecosystems and habitats, such as the once diverse and abundant salmon runs, that initially drew colonists. Nevertheless, people continue to move to Seattle, drawn by newly abundant tech-sector jobs and the beautiful landscape. The city's policy legacy reflects the value that contemporary residents place on the environment. By the 1970s, Seattle's city government was investing in tree-planting and bike lanes and promoting voluntary composting.

The environmental sensitivity exhibited in certain city programs, though, did not protect Seattle from a garbage crisis in the 1980s. Nor did it initially influence how the city framed its response to that crisis. Seattle leased two landfills (shown on the map in figure 3.1) in the neighboring city of Kent for commercial and residential waste, and by the 1970s, both were rapidly filling up. Without any obvious alternatives, waste managers floated all kinds of ideas for redirecting and reducing the waste stream. But they didn't implement anything.

By 1980, both landfills were leaching toxic substances into surrounding

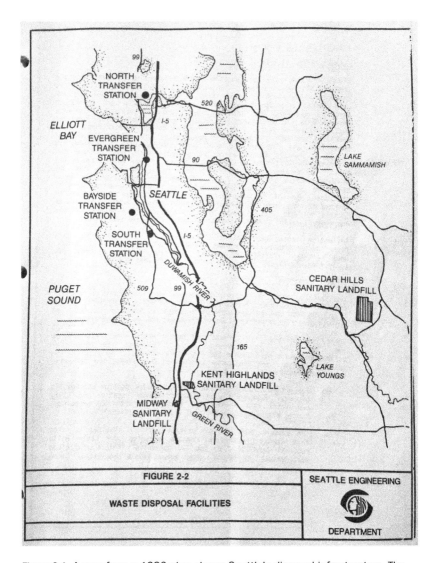

FIGURE 2-2

WASTE DISPOSAL FACILITIES

SEATTLE ENGINEERING

DEPARTMENT

Figure 3.1. A map from a 1986 plan shows Seattle's disposal infrastructure. The Kent landfills (Highlands and Midway) are visible toward the bottom of the frame; King County's Cedar Hills Regional Landfill can be seen to the east of the city. The North and South Transfer Stations remain key parts of the city's disposal infrastructure today. Source: Seattle Solid Waste Management Study—Policy and Development Plan, January 1986, SMA/3127.

air and waterways. They ceased operation in 1983 and 1986, and were then both designated Superfund sites. Unlike Boston, where waste management was financed through property taxes, Seattle charged residents directly for waste collection. The year the first landfill closed, rates spiked as the city scrambled to redirect garbage to the county's landfills. In the face of reduced disposal capacity and increased costs, Seattle's Solid Waste Utility[4] defined its responsibility for waste management in pure Sanitary City terms: "Garbage collection and sewer service," the Utility proclaimed in its 1983 Annual Report, "function to prevent disease and provide for a beautiful environment." And with that, in a mirror of what Boston was doing, Seattle's leaders set about solving a disposal problem.

Seattle ultimately abandoned the disposal-problem frame. Citizen input encouraged city leaders to reconstruct the garbage problem as a resource problem, which eventually led the city to drop plans for a waste-to-energy plant and instead focus on ambitious recycling and waste reduction goals. This transformation reconfigured the traditional roles of citizens and the state in waste governance and management. Citizens took on more responsibility for their discards—but not strictly on terms dictated by the WRWR. Instead they became active partners in municipal waste management by engaging more—not less—with their discards. The city government expanded its role from service provider to consumer, educator, and provider of effective, convenient, and comprehensive infrastructure. These transformations restructured waste management in the city, orienting every decision around resistance to the WRWR.

The Disposal Problem

From the Sanitary City perspective of the Utility's staff, the city's waste system was in crisis because a loss of disposal capacity had caused skyrocketing costs. The solution, therefore, was to create more disposal infrastructure. Just as in Boston, Seattle's engineers favored WTE incineration, or energy recovery, as it was more commonly referred to in Seattle. Just as in Boston, private-sector incinerator consultants made a strong case for the technology.

Utility engineers first highlighted energy recovery in a proposed strategy that was codified in 1978 by City Council Resolution 25872. The resolution established a four-part system for Seattle's waste: landfill, recycling, energy recovery, and yard waste composting. The landfilling and energy recovery portions of the resolution provided detailed guidelines for evaluating options, determining risks, working with neighboring governments, and identifying community and environmental impacts. The resolution was equivocal, however, about the role of recycling and composting in the city's waste system. It directed the Utility to devise multiple scenarios to account for the failure of the city to meet "realistic" recycling and composting goals. The commitment to WTE and landfilling, alongside the sketchy approach to composting and recycling, demonstrated the extent to which the latter were ancillary to the central function of the Utility: to ensure proper waste disposal.[5]

Throughout the early 1980s, the Solid Waste Utility worked within the framework set by Resolution 25872, consistently promoting energy recovery as the obvious path forward. Not only would WTE provide long-term and stable disposal capacity, it would also create energy. This was a nontrivial benefit, as the city's electric utility, City Light, was still recovering from an acute energy crisis at the end of the 1970s.

City officials viewed energy recovery as a long-term solution, but one with a long lead time. While planning was underway, the Utility did make efforts on recycling and yard waste composting. In internal documents, these efforts were framed as a way to reduce the city's "disposal burden,"[6] which cast these experimental programs in WRWR-sanctioned terms. The language used publicly seemed designed to resonate with Seattle's ecologically conscious constituents, with program names such as "Use It Again, Seattle."[7] In practice, though, the programs simply outsourced waste reduction efforts to residents, using neoliberal strategies acceptable to the waste regime. Sam Sperry, then head of the city's Engineering Department, parroted the WRWR logic perfectly in his justification for the program: "to reduce City involvement in solid waste management, to increase citizen control over their own affairs, and to maximize consumer savings in solid waste, management and gardening costs and to save energy."[8]

There was a loose consensus at the time, likely magnified by ascendant Keep America Beautiful messaging, that the city should be working actively to turn more control for waste management over to citizens—or at least to occupants of single-family homes with backyards. Sperry did note at the time that "a centralized project would be necessary to compost all yard waste," and that the city's research surrounding composting revealed that it would be "more complex and more intensive than originally planned."[9] Ultimately, the efforts launched by the city in 1980 nominally followed the guidelines from Resolution 25872, and did so without requiring any substantial investment from the city. The city's more capital-intensive efforts went toward planning and preparing for new disposal capacity.

The Utility's minimal efforts toward waste diversion and citizen composting reinforced WRWR logic and the city's commitment to resolving a "disposal problem." This commitment was reinforced again in the Utility's 1981 "Proposed Recycling and Waste Reduction Strategy," which aimed "to reduce the amount of waste requiring disposal through composting, recycling and other waste reduction methods; and to recover energy from the remaining waste, with land disposal of residuals."[10] Although the problem definition was the same, we already see some key differences from Boston, likely owing to Seattle's larger engineering staff and Utility structure. Even early in the planning for energy recovery, and fully within the WRWR framework, waste managers in Seattle understood energy recovery as part of a larger system of consumption and disposal. But the basic logic propelling the two cities to energy recovery was the same. Even though Seattle's plan included "waste reduction" in the title, it was oriented to end-of-pipe solutions. Like the "Use It Again" composting program launched the year before, it included minimal investment in non-disposal infrastructure and programs. The plan sought to capitalize on the fact that Seattle residents already recycled through a network of private recyclers without the city government having to make any direct investments.[11] In short, with energy recovery still years away, the Utility was committed to solving a disposal problem, and officials — with broad citizen support — relied on low-cost diversion programs designed to ease the city's disposal burden.

In public discourse, local newspapers normalized and reinforced the disposal-problem framing that the Utility had crafted and that formed the basis of its planning. Reporters would almost uniformly open articles with statistics about the growing volume of Seattle's waste and its shrinking landfill capacity. Through their language, reporters reinforced the idea that the Utility's primary responsibility was disposal. One particularly hardboiled account put it this way: "For years this 'out of sight, out of mind' approach to throwaways worked well, but in the past decade the Engineering Department has discovered that yesterday's paper is still bad news. Seattle's landfills [are] filling up faster than you can crumple a beer can and toss it in the rubbish."[12] Though everyone acknowledged that waste volumes were growing, the popular framing reinforced the understanding that the problem that needed solving was a lack of disposal capacity.

Reconstructing the "Disposal Problem"

Seattle was not at all unique in initially framing its crisis as a disposal problem. At the time, the same definitions and assumptions dominated in Boston and in countless other cities across the country. But as planning for energy recovery proceeded, Seattle redefined garbage as both a *thing* and a *problem*, ultimately challenging assumptions of spatiality, materiality, and temporality as constructed within the WRWR. This redefinition launched Seattle down a path of regime resistance.

As city officials began to recognize the potential in both energy recovery and recycling, they began to see a new identity of waste as a resource, amplifying age-old American values of thrift and economy. As the politics of disposal within the region came into focus, however, waste was redefined again, and more meaningfully, as a source of political power, with county and city leadership competing for control over it. Finally, through a process of reckoning with the potential consequences of energy recovery, residents, politicians, and waste managers began to explore the material and temporal life of garbage, giving character and a future to what was once understood in blunt, abstract, atemporal, and strictly negative terms. New waste management possibilities emerged from these discourses that challenged the WRWR.

The first key shift in the city's definitions occurred as part of the energy recovery discourse. Although it was a solution born from Sanitary City logic, the idea that waste could be converted into a commodity shifted the contours of the disposal problem slightly, even before planning began in earnest. News articles with telling headlines, such as "Who Gets the Garbage? Is There Gold in Those Cans?"[13] and "Put Garbage In, Get Power Out,"[14] introduced to the public the idea that trash could be a potentially valuable resource. As the energy recovery plan gathered momentum, Seattle's mayor, Charles Royer, told a local paper that "burying garbage is not cost effective. . . . Garbage is now worth money."[15] This concept was reinforced by the Utility's financing structure. Theoretically, revenue from energy recovery could eventually offset the solid waste rates charged to customers. Because Seattle did not seriously consider contracting with a privately owned and operated plant, energy recovery could have substantially altered the finances of waste management.

The prospect of cost reduction was attractive, but, as the energy recovery project became more concrete, opposition to the project also crystallized. The potential value in garbage took on another layer of meaning as advocates pushed recycling as a viable alternative to disposal. Whereas energy recovery redefined garbage as another abstraction—a commodity—composting and recycling offered the possibility that discarded materials had value *in themselves*. Having already pressured waste managers to include recycling and composting in the city's initial 1978 waste management strategy, residents continued to clamor for aggressive recycling to be considered alongside more traditional disposal options. Recyclable materials, residents and activists argued, were resources that shouldn't be wasted.

Constituent pressure worked. By 1984, energy recovery had shifted from being seen as the primary solution to being seen as a secondary solution — the solution for the residual waste *after* recycling had already taken place. The city council was considering a recycling goal of 40 percent of the waste by tonnage, with energy recovery only for the remaining portion.[16] No city in the country had a recycling rate even remotely close to 40 percent at the time. Seattle's leaders were willing to consider novel possibilities.[17]

Citizen opposition to incineration was accompanied by pressure of another sort. Constructing an incinerator would have required cooperation, in one form or another, with King County. As Anna Davies, Zsuzsa Gille, and others have observed, garbage is a geospatial phenomenon.[18] It both defines and is defined by the places where it exists, and this feature of waste can be exceptionally problematic when it shifts across scales or boundaries of governance. In Seattle, negotiations about what regional "cooperation" might look like gave rise to a new geopolitical identity for garbage: a source of power in the region.

As Seattle grappled with what do in the wake of its landfill closures, it began sending its waste to King County's Cedar Hills Regional Landfill. The arrangements with King County were tense and expensive, and Seattle viewed Cedar Hills only as a stopgap measure. In order to proceed, though, the city had a number of difficult decisions to make. The key question was whether the city would hand its waste over to the county to manage, partner with the county on a regionally coordinated system, or go it alone. The city viewed this question as historically unique and terribly important to the city's future.[19]

Initially, Seattle's leaders and many constituents were in agreement that proceeding in partnership with the county made sense — for many reasons, including the fact that regional coordination was required by state law and necessary for EPA planning funding.[20] In 1986, after years of consideration, Mayor Royer formally declared his intention to partner with King County on the development of energy recovery. He argued that the city could "no longer go it alone." Seattle needed the county landfill until the energy recovery plant could be completed. Furthermore, the county might require a regional network of energy recovery facilities, which would require close regional coordination. Finally, Seattle constituents "overwhelmingly favored" a regional approach. "Our citizens," the mayor argued, "and the beautiful but fragile environment we all cherish, deserve no less" than "the best regional solid waste system in the country."[21]

With the reality of incinerators on the table, the mayor hinted that a regional approach had the potential to reduce the city's responsibility for waste infrastructure. As Engineering Department director Sperry had said earlier, "Regional planning, which is likely to look at more options than a City-only project, is likely to benefit City residents."[22] The popular support for such a measure suggests that many in the city hoped that the county might host the undesirable infrastructure while accommodating Seattle's discards. This is the standard spatial politics of the Sanitary City: elsewhere is always a better place for waste.

By mid-decade, however, several city councillors had begun to sour on the idea of a regional partnership. In September 1986, city council staff recommended that Seattle should maintain its independence.[23] The council's Environmental Management Committee unanimously supported the recommendations of the Utility's 1986 solid waste plan—that is, the "Solid Waste Management Study: Policy and Development Plan" (SWMS, vol. 2)—with the exception of the issue of who would control the waste stream. The city council staff argued that regardless of whether the city or the county controlled the waste stream, energy recovery was likely to be part of the solution. The key was for the city to ensure that Seattle was able to maintain "a voice in decisions that affect City ratepayers."[24]

Utility staff also began to worry about a regional partnership. "There is no clear advantage," Utility staff argued to the mayor, "for either the City or the region in the City's relinquishing its responsibility for waste management to the County or a regional authority."[25] Exporting waste and ceding disposal responsibility would reduce the city's ability to control costs. Furthermore, given the size of Seattle's waste stream, combining the city's waste stream with the county's would still likely result in an incinerator within Seattle's borders. And because Seattle, at the insistence of residents, intended to accompany incineration with an aggressive recycling program, the Utility anticipated slower waste stream growth in the city than in the county. A likely outcome of partnership would therefore be a Seattle-based facility that would have to accept county garbage in order to operate at peak efficiency. The worst outcome, in the minds of city leaders, would be a city facility that had to burn non-city waste. Utility staff argued to the mayor and the city council that Seattle could build a smaller facility and maintain more control over waste reduction efforts if it moved ahead on its own, independently from the county.[26]

Responding to the Utility's arguments and city council concerns, the mayor reconsidered his position. In late 1987 he acknowledged that "Seattle's waste poses special problems for the region: We simply have too much waste to be accommodated in someone else's disposal system." But to soften the politics, the mayor argued that "Seattle will behave most responsibly in partnership with the rest of the region not by dumping our responsibilities in someone else's lap, but by developing our own system in a tightly coordinated, well-planned regional system of waste management."[27] By opting to maintain independence, Mayor Royer positioned Seattle, and himself, as a responsible actor in the region. At first a material that no one wanted, garbage now had the ability to confer power and political legitimacy, not to mention potential revenue, to whoever controlled it.

It is worth noting that, within this regional powerscape, one of the city's most promising independent incinerator options initially was from the Tulalip tribes, who proposed to locate an incinerator for Seattle's garbage on reservation lands outside of Maryville, Washington. The tribes understood that controlling the city's garbage would be lucrative. The mayor explained that there might be "an advantage of locating a facility in a receptive community," but that there were also disadvantages to the proposal, including transportation costs and loss of control.[28] He was silent on the complicated power dynamics of ceding that control to a tribal organization that would then also have the most immediate exposure to any pollution from the plant.

As the regional politics came into focus, local and regional advocacy organizations and Seattle residents—who had once strongly supported a regional solution—also began to advocate more actively for an independent system with local facilities. One constituent was particularly direct on this point, telling the mayor, "I believe strongly that the city should take responsibility for its own garbage, rather than passing that responsibility off to the county. We create it, we should deal with it."[29] In similar, if more diplomatic, terms, the League of Women Voters and the Municipal League of Seattle/King County released a joint statement arguing that "as the producer of roughly one-third of total county garbage it remains Seattle's responsibility to develop an appropriate disposal system for a corresponding amount of solid waste."[30]

King County, meanwhile, was not idly watching the process unfold in the city. In August 1987, just as the mayor was deciding to withdraw from a regional partnership, the county proposed new rates for the Cedar Hills Regional Landfill. The new rates included a ten-dollar per ton surcharge for Seattle. The county justified the charge as protection of the limited capacity and an incentive for Seattle to proceed with planning alternatives, but the mayor interpreted it as an unjustified penalty.[31] Immediately following the announcement of the new Cedar Hills rates, Tim Hill, the county executive, also informed Mayor Royer that the city would need a county permit to build an incinerator in Seattle; the letter suggests, though does not state outright, that the city should not assume that such a permit would be easily forthcoming.[32] A few months later, Hill gave Seattle an ultimatum: stop using the county's facilities within five years, or commit to using them for forty more years.[33] County leadership recognized that control of Seattle's waste was a form of political power, an understanding that was only reinforced by recognition of the potential financial value of controlling the city's garbage. Indeed, county management of the city's waste, particularly through energy recovery, could have been profitable, and it could have locked Seattle into a dependent relationship with the county for decades to come.

The prospect of energy recovery prompted politicians to consider garbage as a commodity, a shift that then transformed the regional politics of waste. Controlling Seattle's trash morphed from a "burden" to a responsibility that conferred power. Once waste was recognized as a source of value and power, planning momentum shifted. A broad spectrum of participants contributed increasingly radical ideas about managing discards and stewarding resources that resulted in new material and temporal identities for garbage in Seattle. These new definitions negated the legitimacy of energy recovery as a solution altogether.

Seattle was not unique in its residents' interest in waste management, but it was unusual in the capacity of local government to receive resident feedback and translate it into tools for action. The broad discourse among city and citizen stakeholders resulted in new material and temporal identities for garbage in Seattle. By the 1970s, Seattle had developed robust public input processes, and citizen participation was an integral part of all planning and policy efforts in the city. The city government, both within and beyond the Solid Waste Utility, was organized to receive and absorb public input. Although these features have not historically been inclusive of all voices,[34] the institutions of public participation were relatively strong, especially in comparison to cities in other parts of the country.

Three public-facing organizations within city government influenced waste planning: the Solid Waste Utility, the mayor's office, and the city council. Representatives from each of these bodies actively engaged with Seattle residents throughout the 1980s through both formal and informal channels. Informally, city officials received, responded to, and shared letters from residents. Thoughtful and individualized responses to citizen inquiries from Diana Gale, head of the Solid Waste Utility from 1987 onward; from Virginia Galle, head of the city council's Environmental Management Committee; and from Mayor Royer's office suggested a culture of responsiveness in city government.

Citizens were, at first, most interested in the frequent and dramatic rate hikes over the decade. In 1980, the city council approved a variable rate for garbage. For the first time, Seattleites were charged for waste collection based on the amount of trash they set out rather than a flat fee. The Utility undertook a thorough study to determine how to set the new rate structure in 1979,[35] but was forced, almost immediately, to raise rates. Subsequently, the Utility raised rates almost annually, a practice that continued throughout the 1980s. The regular rate hikes made Seattle residents keenly aware of their garbage and more invested in identifying a sensible — and less expensive — solution. The monthly bills provided a channel through which the Utility could communicate with residents and a hook that kept residents connected to the ongoing crisis. Rates rose particularly sharply in 1986, after both of Seattle's last dumps in Kent were labeled Superfund sites that would require very expensive remediation.[36] This rate hike alerted residents to the expensive future of their past discards. As the decade wore on, alongside their complaints about rates, residents increasingly wrote to city officials about new programs and plans. City officials received thousands of letters from constituents about solid waste planning over the 1980s.

On top of this consistent, informal exchange among residents, solid waste managers, and elected officials, there were also formal input processes structured into planning and decision-making. A citizen's Solid Waste Advisory Committee was convened in the late 1970s to advise on the creation of the city's 1978 solid waste strategy. Unlike Boston's narrowly mandated equivalent, this committee, which consisted of resident, nonprofit, and business-sector representatives, remained functional after completing its initial charge, formally advising the city in all its planning and policy decisions concerning solid waste. The committee still exists, and any Seattle resident may apply to serve on it.

The Solid Waste Utility produced dozens of planning and policy documents from the late 1970s through the 1980s, and each of them had its own participation and comment process. A few of the most significant included the Recycling and Waste Reduction Strategy, a composting study, a series of rate studies, and a series of comprehensive plans. These were followed up with the SWMS Vol. 2 in 1986, which ultimately served as the guiding policy document for the rest of the decade. The process for generating the SWMS incorporated a broad set of professionals, including experts in communication, engineering, ecology, public health, and solid waste infra-structure. The process also included public involvement, from the earliest stages of determining the scope through the selection of preferred options. The final document included many recommendations that had come directly from Seattle residents.[37] The SWMS and most of the other plans were pre-pared with traditional consultative public input, through public meetings, and also were presented in public hearings to the city council.

An environmental impact statement (EIS) prepared for the energy re-covery project was the most significant planning document from the decade, in terms of both its influence and its public input. From the very beginning, city officials were committed to conducting a complete and thorough re-view of all of the city's options. In a newsletter about energy recovery that went out to ratepayers, the Utility announced, "To make sure that the draft EIS for the Energy Recovery Project is as complete as it should be, and, in fact, asks all of the important questions, deciding the scope of the draft EIS for this project will be highlighted at a series of community meetings." The newsletter went on to say that the meetings were "planned to give people in Seattle a forum for asking questions and voicing their comments and con-cerns about energy recovery" and to assure residents that "the issues raised in these 'brainstorming' sessions will then be incorporated into the draft EIS."[38]

The solid waste planners were quite true to their intentions to use citizen input to shape the EIS from scoping through option evaluation. As a re-sult, and in stark contrast to Boston's limited EIR, Seattle's EIS was given a broad scope and considered a range of options, from energy recovery alone to an aggressive recycling and waste reduction program with landfilling of residuals. Throughout the EIS process, the Utility and its consultants re-mained open to formal and informal public input. During the preparation of the first draft EIS, it received dozens of written comments, not to men-tion several hours' worth of verbal testimony at meetings, all of which was recorded and which the Utility responded to.[39]

Three issues dominated public conversation as the EIS planning un-folded. The first of these was the remediation of the two Kent landfills; the previous generation's disposal solution had become a pressing, smelly, ugly, and expensive burden for current and future Seattleites. The long-term health and environmental risks associated with incineration constituted a second factor. Residents framed their concern about these uncertain risks as imposing unfair and unnecessary burdens on future generations. Third, as residents weighed the pros and cons of a possible incinerator in their midst, they focused on the specific material composition of the waste stream, emphasizing the opportunity cost of burning organics and other materials that were easily recyclable. When given space to participate meaningfully in the waste planning process, residents engaged with the material and temporal life of garbage in ways that contrasted with and challenged the definition of waste in the WRWR. Residents refused to accept waste as an abstraction that belonged elsewhere—or that somehow "disappeared"—because the things they had thrown away before were already imposing concrete impacts on their spaces and bodies. These impacts were likely to continue, possibly in new and different ways, if the city built an incinerator.

The deterioration of the Kent landfills further emphasized the material presence of Seattle's waste. The landfills were frequently described in local papers in sensory terms. Most commonly, reporters focused on their olfactory impacts. But the dumps were also a visual presence. A neighbor of one of the Kent dumps told a local reporter, "It's been smelly for years and it looks like a junk shop."[40] One particularly big rainstorm in 1982 left a stinking, toxic, anaerobic, forty-foot-deep pool in one of the landfills that neighbors christened "the Black Lagoon."[41] "Stories of neighborhood children wearing protective goggles, hundreds of complaints, and a threatened lawsuit by Kent officials," one paper reported, led the mayor to announce that the second landfill would close a year and a half ahead of schedule.[42]

After both Kent landfills were closed, Seattle had to confront the fact that not only was its waste not "away," but it wouldn't stay in its place. Smells, leachate, and methane migrated beyond the borders of the landfills even after they stopped accepting new waste. The *Seattle Times* reported that "Bobbie Gojenola, who lives just to the west, said she still will worry about water draining into the water table from the dump and flowing toward the nearby elementary school."[43] Gojenola's worries, and those of countless other landfill neighbors, were only magnified by chronic and persistent methane leaks at both landfills.

As very real sights and smells emanated from the closed landfills, some constituents began to imagine the physical experience of breathing Seattle's incinerated waste. In a scenic postcard of downtown Seattle under fluffy, cumulous clouds, a resident contrasted Seattle's tourist-friendly image as a clean and beautiful city with the oppressive reality of urban air pollution (figure 3.2). "What will happen," she lamented, "when we add smoke from a garbage incinerator to the smog?" For Seattle residents, contaminated water, highly combustible methane gas, and smog all embodied the ongoing material life of Seattle's garbage.[44]

In 1986, as the astronomical costs of remediating the Kent landfills inflated residents' garbage bills, Mayor Royer sent a letter to the city council

SEATTLE, WASHINGTON

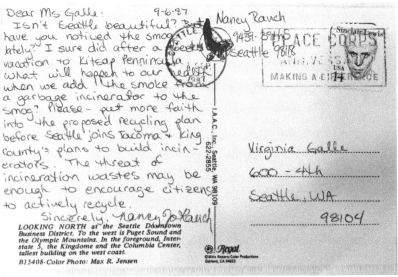

Dear Ms Galle: 9-6-87   Nancy Rauch
   Isn't Seattle beautiful? But
have you noticed the smog   9451-59th S
lately? I sure did after a week   Seattle 98118
vacation to Kitsap Penninsula.
what will happen to our health
when we add the smoke from
a garbage incinerator to the
smog? Please - put more faith
into the proposed recycling plan
before Seattle joins Tacoma + King
county's plans to build incin-   Virginia Galle
erators. The threat of   600 - 4th
incineration wastes may be   Seattle, WA
enough to encourage citizens
to actively recycle.   98104
         Sincerely, Nancy Jo Rauch

LOOKING NORTH at the Seattle Downtown
Business District. To the west is Puget Sound and
the Olympic Mountains. In the foreground, Inter-
state 5, the Kingdome and the Columbia Center,
tallest building on the west coast.
B15408-Color Photo: Max R. Jensen

I.A.A.C. Inc. Seattle, WA 98109 622-2855

Regal
© Mike Roberts Color Productions
Oakland, CA 94623

Figure 3.2. A postcard from Seattle resident Nancy Jo Rauch to City Councillor Virginia Galle, September 6, 1987, urging recycling instead of incineration. Source: SMA/4630-02/7-13.

president that indicated a new respect for the spatiality, materiality, and temporality of trash. It was time, the mayor argued, "to move into the next era of solid waste management": "For more than a half century," he added, "we have simply buried our solid waste in the ground, first in Seattle, and when there was no more vacant land here, in the city of Kent. The legacy of those landfills has now become one of our worst problems."[45] The mayor recognized that moving waste farther away in space had not prevented it from transmuting into other kinds of real, material problems. He positioned the current moment between a history of shortsighted waste management decisions and a future burdened with the consequences.

While the ongoing saga of the Kent landfills drew attention to the physical presence of garbage, the most promising solution—energy recovery—posed new material and temporal concerns for many constituents. Some residents reacted in classic NIMBY fashion, recommending that the city revert to a regional approach based on waste export (to "away") rather than building an incinerator in Seattle.[46] But many, many more raised substantive concerns and offered viable alternatives. Many of these constituents worried that elected officials and city staff were not thinking beyond the immediate crisis. City officials initially argued that the incinerator option was a long-term solution to the landfill crisis, but many constituents disagreed. Seattle residents argued that the expediency of the incinerator option was at odds with its long-term environmental consequences. As one constituent rather dramatically put it, Seattle's decision would "impact the ecological balance of our planet for all time. I urge you to vote against incineration of our garbage. We can no longer afford short term solutions to long term problems. . . . [T]here are other methods of dealing with garbage[.] [P]lease take the extra time to explore all possible solutions."[47]

Residents further critiqued incineration as a toxic folly that would materially harm already overburdened communities. Once the city decided to move ahead independently from the county, it identified a set of potential sites in the city. These sites were located within an industrial belt at the center of the city that was also home to much of Seattle's working-class population. Many residents of neighboring Beacon Hill wrote to the city council and mayor in strong opposition to the incinerator, arguing, for example, that the area was "unhealthy enough" and did "not need any more pollution and odor from such a plant."[48] Locating the facility in this area was expedient from a zoning and land use perspective, but would have exacerbated existing environmental injustice. Concentrating polluting facilities in areas zoned for such facilities only reinforces historic spatial patterns of unequal exposure and risk.[49] In Seattle, the decision would likely have depressed land values further while also generating a variety of nuisances and pollutants. Instead, residents of Seattle's industrial areas urged the city to institute recycling programs "with teeth."[50]

As public dialogue about incinerator impacts began to clarify the temporal-material identity of garbage, residents began to advocate vociferously for recycling and source reduction over disposal. In so doing, they chipped away at the disposal paradigm of the WRWR, carving a monolithic, atemporal abstraction into its constituent parts. City policy-makers and waste managers began to understand waste as a set of material resources for present and future generations.

Energy recovery, residents argued with increasing urgency (echoing the writer of the postcard in figure 3.2), was materially wasteful. Recycling was a more sensible solution for many, if not most, of the constituent materials of the waste stream. "A mass burn incinerator is an attempt to deal with symptoms," one resident wrote to the mayor, "but is not a solution to the real problem of waste management. Incineration can further degrade the air quality of Seattle, release toxic substances into the atmosphere, and create new disposal problems with hazardous ash. . . . Waste reduction thru recycling should be aggressively pursued before planning and designing any incineration plants."[51]

A flyer developed by a residents' organization in south Seattle, where the incinerator was likely to be located, cast energy recovery as a wasteful threat: "Because energy can be generated by burning garbage, incineration has been given the name 'energy recovery,'" the flyer stated. "But don't let the name fool you." It further explained that incineration was less efficient than waste reduction or recycling if the goal was resource conservation. "In fact," the flyer noted, "incineration can actually pose a threat to conservation, for once an incinerator is built, it creates a demand for a steady supply of waste, much of which could be recycled and eliminated from the waste stream."[52]

The argument for the conversion of an abstract waste stream into a stream of reusable materials built on the dialogue that had accompanied the development of Seattle's composting program, which had focused on isolating yard waste from the waste stream and reclassifying it as a resource for urban gardeners.[53] Likewise, the city's preliminary waste reduction and recycling plan from 1981 had identified easily recyclable materials that were ending up in landfills, including paper, metals, and glass.[54] In 1980, Jim McMahon, a consultant to the Solid Waste Utility, explicitly framed waste as a potential, if nondescript, resource: "The point is to get the mechanism set up for the long-term recovery of resources . . . such that when the day comes[—]and it's coming—when we have to recover our resources, we'll be ready instead of just starting to plan."[55] Through half a decade of planning efforts, certain materials in the waste stream had already been identified and reclassified; but these materials, and waste in general, were still being abstracted into large categories, with little, if any, exploration of the unique characteristics or future uses of each one.

Building on the strong recycling ethic in Seattle and the recycling and composting planning the Utility had done in the late 1970s and early 1980s, by the mid-1980s the city was considering recycling as a cornerstone of its waste management strategy. From the mid-1980s on, the Utility and Mayor Royer, motivated by citizen recommendations, explicitly argued for an incinerator sized to accommodate Seattle's residential and commercial waste streams *after* recyclables had been removed.[56] Toward the end of 1987, reflecting growing popular support for recycling, the mayor formally recommended pursuing the "Reliance on Recycling" alternative from the 1986 SWMS, which called for recycling 40 percent of Seattle's waste by 2010.[57] Though there was no precedent for a 40 percent recycling rate in any US city at the time, Utility officials and elected leaders recognized that residents preferred this direction as a matter of policy. Furthermore, any material that was recycled sidestepped the complex regional politics discussed above—and if the city did proceed with a county partnership, then aggressive recycling would diminish the power the county wielded over the city.[58]

Once the 40 percent goal was established, residents from across the city audaciously contended that it was too low, particularly in combination with investment in energy recovery. The city was still on a path to waste millions of dollars in addition to the material resources to be burned. One letter, addressed to City Councillor Virginia Galle, the head of the council's Environmental Management Committee, admonished:

> Put your money where your mouth is. The city is planning on spending up to $200 million on an incineration plant while spending roughly $1.5 million for recycling each year. Joe Garbarino, private owner of the largest recycling facility in the country, estimated that Seattle could recycle fifty percent of its waste stream within five years if it would make the capital investments necessary to separate relevant materials from the waste stream. Not ten years, not twenty years, five! Even with the proper capital investment the city would spend only a fraction of what it would for a burn plant. If the solid waste utility put half the time, money and energy into recycling that it has put into greasing the tracks for incineration, we would see immediate and drastic improvements in waste reduction.[59]

Other citizens explicitly argued that the city was in danger of misusing resources. "I would like to see all recyclables diverted from the waste stream and recycled," wrote one resident to the mayor. The author of the letter argued that incineration was a "misuse of resources" and said, "I realize it will require a great deal of public education, but I would like to see the city set its long term goals for recycling higher than the proposed 40 percent."[60]

It is possible to interpret these citizen calls for recycling as evidence that Seattle residents had internalized neoliberal weak recycling waste regime messaging about individual action, and were simply amplifying corporate talking points designed to obscure corporate responsibility. But while it is true that recycling can provide cover for increased production of disposable consumer goods, in the context of the incineration debate the residents were actually proposing something more radical.

Waste-to-energy incineration also provides cover for disposability. Incinerator proponents intentionally framed WTE as a form of reuse, arguing that the potential energy in discarded materials would be captured and used, rather than needlessly wasted in a landfill.[61] This view represented limited engineering logic and resonated with corporate waste managers and producers. Recycling and WTE could both be distorted to serve corporate interests. But by resisting incineration and its attendant logics, while demanding that the city invest aggressively in recycling in the context of a long view of material life cycles, Seattle residents resisted the simple logics of the WRWR.

First, citizens were asking for more, rather than less, contact with their discards. Generations of Sanitary City and WRWR waste management had sought to remove waste as quickly as possible, with as little human contact as possible. Modern waste systems were designed to make disposal easy, instinctive, even thoughtless. It is crucial to regime maintenance that people do not think too much about what they throw away. Even in the WRWR, recycling was designed to be basically akin to disposal—just plop a thing in the right bin. By rejecting incineration, which they could have accepted as a form of recycling, Seattle residents were imploring the city to build a system that forced them to increase the time and thought they gave to their discards.

Second, Seattle residents were demanding that city decision-makers consider an unproven policy track that was based mostly on the "nonexpert" knowledge of residents (i.e., about their ability to participate in the process of material sorting at the household level). Moreover, they were suggesting that the city itself needed to continue to play a significant role in waste management, though a role that was different from what it was accustomed to: it had to be not only a service provider, but also an educator. At a historical moment when the private sector was assuming increasing control of waste processing and disposal, the act of demanding more and deeper engagement from the city government was also a form of resistance. Municipal recycling of packaging materials was certainly from the KAB playbook, and corporate actors on the production side promoted it to deflect regulatory attention upstream. But as proposed by Seattle residents, recycling resisted the pull of corporate interests on the disposal side; it also proposed restructured relations at the urban scale that would fundamentally alter the habitual roles of citizen and state within the WRWR. Though this was not a radical break with the regime—most proposals (though not all) were still end-of-pipe solutions and avoided implicating corporate producers—the proposals of Seattle residents represented incremental resistance, the seeds of new material and political relations in waste management.

As residents advanced arguments for a new role for themselves in the waste system, some actually flipped the narrative of the waste problem completely, arguing that incinerators—rather than the garbage—were the problem. For instance, one resident argued explicitly that "incinerators are problems—air quality, the release of extremely toxic fumes, the problem of disposal of the very hazardous ash are examples. . . . Please emphasize recycling in the planning for incinerators, [and] require recycling."[62]

Two other citizens argued that the incinerator was a "shotgun approach" and that the city could instead view garbage as an "opportunity to enhance quality of life." Like many other members of the public, they emphasized specific actions the city could take, including instituting a ban on nonrecyclable packaging, requiring increased municipal recycling, creating incentives for composting, and establishing a tax on disposable goods equal to the cost of landfilling them at the end of their life. These suggestions implicated production and moved beyond WRWR-sanctioned end-of-pipe solutions.[63]

Seattle residents maintained that garbage could be something other than a burden. A thoughtful combination of municipal investment and individual action—a recipe that chafed against simplistic WRWR relations— they suggested, could yield a better solution. These two threads challenged the regime definition of waste itself as well as the role of city government in waste management. If waste was a potential resource with immediate and long-term potential to increase the well-being of the city, then the role of city government was as a steward rather than a "manager." This perspective implies a whole different set of tools and obligations: city leaders should be focused on shifting incentives, providing convenient options for material recovery, and educating residents and businesses rather than focusing on hygienic removal and disposal.

By the end of the 1980s, evidence of a straightforward "disposal problem" was nowhere to be found in city documents. The city's 1989 solid waste plan prepared by the Utility, titled "The Road to Recovery," described the problem of waste as "difficult" and "uncertain."[64] In addition to a waste-as-resource framing visible in the conversations about the incinerator and recycling, waste was also understood as a source of political power. Furthermore, garbage itself was understood as a complex stream of multiple, specific materials, each with unique characteristics and future possibilities, and which, if mishandled, could haunt the city for a very long time. In fact, many of the former solutions to the disposal problems were, by the end of the decade, understood as problems in and of themselves. Over a decade, city leaders, waste managers, residents, and a host of "experts" had together redefined the garbage problem.

Redefining Waste Management

The new problem definition required new kinds of solutions. Traditional management techniques rested on assumptions about what waste was and who should be responsible for it that were no longer legitimate in the Seattle context. The city's Solid Waste Utility was evolving alongside the public conversation, morphing over the decade from a traditional sanitary engineering department into a professional organization with a much broader mission and set of skills. Over the course of the 1980s, a series of new hires brought in ideas, expertise, and values that extended far beyond sanitary engineering, opening the Utility up to possibilities it would not have previously considered.

The first key hire was Don Kneass in 1984. Kneass, a veteran of the private recycling industry in Seattle, had been brought in to develop and run the new recycling programs required by Resolution 25872.[65] Before he had been able to get the recycling programs off the ground, however, the Solid Waste Utility found itself mired in a full-fledged public credibility crisis that had been building for some time. Methane was leaking uncontrollably from the two landfills in Kent, and it was found that Seattle officials had known about the leaks for months or even years.[66] Garbage rates were climbing,[67] but the new variable rate structure introduced in 1981 was not having the intended effect of reducing waste generation.[68] Skeptics argued that the Utility was squandering public resources.[69] Following a major re-routing, local papers regularly published reports of uneven collection. The head of solid waste, Gerard Fairbanks, stepped down in 1981 amid rumors that he was asked to resign over the methane leaks at the Kent landfills.[70] The following head, Rich Owings, a longtime Seattle Public Utilities (SPU) employee, oversaw several important solid waste plans, but during his watch Seattle also saw a period of skyrocketing rates and two Superfund designations. He stepped down, exhausted, in 1986.

Mayor Royer selected Diana Gale, a young policy wonk with a PhD in urban planning, to replace Owings. Gale had no particular experience in solid waste management, but she had been the head of the city council's policy staff, so she had both a handle on city politics and the trust of the city councillors. In her own words, she felt that her hiring had been an act of desperation:

> I was two weeks into the job and the deputy mayor was introducing me to [an SPU leadership team], which was a group of, you know, eight engineer men. And they had been very clear with the mayor that they wanted an engineer. So the deputy mayor was introducing me and then there was a silence, a pause. And [someone] looked at the deputy mayor and said, "Why did you pick her?" At which point, without thinking, the deputy mayor said, "Well we asked sixty-five people and she was the only one interested."
>
> So ... Well, (a) I don't know if they asked sixty-five people and I was the only one interested, or (b) my thinking was, from the mayor's point of view, this was an impossible job. And that picking me, who was the darling of the city council, would take some pressure off of him. Because the city council would be slow to be terribly critical of me because they liked me. And so I think that was part of his strategy. I don't think anybody thought it could be done.[71]

Gale took the reins of the Utility in January 1987. She inherited the controversial incinerator project, a terribly costly contract with a King County landfill, and a lot of public distrust. In the years prior, the Utility had created one storm after another, including sky-high rate increases, bungled route and service changes, and, of course, the Superfund designations. Despite the mess, the Utility had been listening, and Gale was told during her first week on the job that she needed to get a citywide curbside recycling program up and running in six months.

Gale's first moves were to request a huge budget for communication and to hire a project manager for ongoing waste planning initiatives as well as a new finance person. The finance person was a competent insider, a long-time SPU employee. The project manager was Tom Tierney, a bright public servant who, like Gale, had no background in solid waste. Gale and Tierney, working together with Don Kneass, who finally had a more receptive leadership team, introduced new perspectives and values into the solid waste discourse and amplified ideas that had been present, but not central, to previous eras of planning. Their values manifested in concrete tools and decision-making support structures that nurtured the new understandings of waste coming from Seattle's residents.

Central among these tools were new methods for conducting tried-and-true cost-benefit analyses. Throughout the early 1980s, the Solid Waste Utility conducted a variety of such analyses in its exploration of energy recovery. These analyses provided "the basis for two essential comparisons: costs of building and operating an energy recovery plant with the costs of continued landfilling and the costs of different size energy recovery plants."[72] In 1984, the Utility had proposed a new analytical method, which the mayor ultimately recommended to the city council. The environmental allowance method created "a methodology for making environmental impacts a part of the cost/benefit analyses of solid waste programs."[73]

In 1987, with Gale's encouragement, the city council adopted a resolution allocating a budget for the Environmental Allowance Program (EAP). The intent of the program was to support community-based and private-sector recycling and waste reduction efforts and then evaluate their cost effectiveness against the cost of disposal. In June 1987, the Utility, together with the mayor's citizen Solid Waste Advisory Committee, began to review proposals from a variety of private-sector entities for waste reduction and diversion pilot programs.[74] The EAP provided an opportunity to evaluate nontraditional methods of waste management and funded experimental small-scale pilot programs that would otherwise have been deemed uneconomical.[75]

The economic analysis of the incinerator proceeded in a relatively conventional manner. It was designed to compare disposal alternatives in financial terms.[76] But there were signs that practices like the EAP were starting to influence decision-making. In 1986, the city council staff, then still managed by Diana Gale, recommended that the council "consider funding future recycling programs that are 10 percent higher than the 'avoided costs.'"[77] Unlike a conventional cost-benefit approach, the avoided costs framework, which was detailed in the 1986 Draft EIS, used the current astronomical costs of landfilling at King County's Cedar Hills Regional Landfill as the benchmark. Rather than evaluating each option against all other options to find the cheapest alternative, the city would consider any costs below 10 percent above the landfill costs to be "cost effective." This expanded definition of cost effectiveness allowed the city to consider other values, such as the social and environmental benefits of recycling or composting, when comparing a variety of approaches.

The city council and the mayor supported the Utility's approach. In 1986, the city council's Environmental Management Committee pushed even further by unanimously agreeing that "the City may consider future funding of recycling programs based on long term cost effectiveness." This meant the city would be willing to subsidize start-up costs for programs that would be price competitive with disposal in the long run, but required substantial up-front investment.[78] When Gale took the reins of the Utility, she made full use of this framework. By 1988, the Utility was consistently presenting financial analysis in an avoided-costs framework.

By taking this approach, officials from across city government relegated fiscal costs—usually the driver of municipal decision-making—to a supporting role. They were able to do this because they had widespread public support for alternative waste management methods, because they believed that the future costs of pollution could be substantial, and because they were supported by a variety of professionals who were able to translate the desire for new alternatives into tools and language that both public officials and the public recognized as legitimate.

By the end of the 1980s, the Solid Waste Utility was a different organization than it had been a decade before. In 1980, the Utility was a sanitary engineering department working within Sanitary City logic. It promoted cost effectiveness, efficiency, and hygiene. It offered problems and solutions commensurate with that value set, tweaked just slightly to accommodate the eco-identity of Seattle.

By the end of the decade, the core functions of the Utility had shifted. During a period when many city governments were shedding staff, privatizing infrastructure, and doubling down on techno-managerial and neoliberal approaches to infrastructure governance, Seattle was doing the opposite. Although the Utility continued to dispose of the city's waste, it was centrally focused on communication, planning, research, and resource stewardship. The staff consisted of communications specialists, economists, ecologists, and urban planners alongside more traditional waste managers. This new set of professionals brought new ideas, and was able to convert ideas from residents into concrete and achievable proposals. As these shifts took place, the basic relations established during the Sanitary City and reinforced by the WRWR gave way to something new.

The Sanitary City waste system revolutionized municipal waste management, transforming it from an individual or self-organized responsibility to a publicly managed municipal service. Within the Sanitary City waste regime, citizens were clients, the government the service provider. The WRWR rearranged these dynamics slightly: within the WRWR, citizens were waste generators, responsible for buying stuff and then discarding it in the right place, and the city was the manager, charged with getting rid of everything that was considered garbage.[79] By challenging the legitimacy of energy recovery and demanding a more active role in waste management policy and practice, Seattle's residents challenged their designated role in the WRWR. By adopting a more proactive stance, residents pushed the city government to shift from being just a service provider and technocratic manager to becoming a facilitator and a resource steward. Seattle city government responded by providing services and infrastructure, cultivating channels for material reuse and recycling, and actively working to reshape the perceptions, knowledge, and practices of citizens and businesses. In some ways, these moves were consistent with neoliberal WRWR structures, but they also forced residents to have more awareness of materials and more contact with their waste. These competencies then became the foundation for more radical transformations later on.

Through the transformation of solid waste management in the city, citizens, the Utility, and elected officials coalesced around the idea that residents should have more responsibility, and that the city's role should be expanded beyond service provision. These shifts appeared in debates in many ways. One way was the consistent public recognition of the "citizens' resolve to recycle."[80] The conventional wisdom in the recycling community at the time was that recycling behavior was a culture that could be cultivated. As one recycling professional from New York City told the annual Recycling Congress in Seattle in 1986, "Punitive measures are perfectly valid when the goal is to protect the public good, but peer pressure is a much more effective measure."[81] The city government in Seattle responded to this context by working to determine how city services and infrastructure could nurture and expand existing citizen commitment to recycling.

By the start of the 1980s, city residents were, by all accounts, good recyclers, a fact that convinced many leaders that the residents would be able to bear increased responsibility for waste sorting in the home. By 1978, the Utility estimated that 14 percent of the city's waste stream was being independently recycled. By the early 1980s, this figure had climbed to 20 percent, far higher than almost any other US city.[82] This was all done without support from the city government. The robust private recycling industry, and the fact that residents had proven they were capable of diverting significant amounts of material from the landfill on their own, contributed to the sense that individuals could and should play an active role in managing the city's discards.

The Utility and the mayor believed in residents' capacity, but, at least at first, they resisted stronger municipal involvement. Throughout the 1980s, city leaders maintained a strong conviction that recycling should be voluntary. This attitude was at least in part a legacy of previous planning efforts, and perhaps part of the hands-off ideology exhibited in the original 1980 composting program. It was also a crucial piece of the WRWR, which encouraged recycling to salve consciences, but actively discouraged the construction of a robust system of material reuse and reprocessing. The city's 1981 Recycling and Waste Reduction Plan was strongly oriented to providing education to encourage recycling behaviors. But even as the city's focus shifted away from energy recovery and toward recycling and waste reduction, the emphasis on voluntary programs remained. By the mid-1980s, when the city council, Mayor Royer, and the Utility cohered around a 40 percent recycling goal, the mayor began to characterize the role of the city thus: "If we are to attain our goal of 40 percent level of recycling, we must change our present collection system to make recycling more convenient for the City's households."[83] The idea was that if the programs were easy enough, residents would participate sufficiently to meet ambitious goals.

Even though the emphasis on voluntary citizen participation remained consistent, there was a distinct evolution in the role the city was playing. The mayor spoke often of "accommodating" the recycling demands of Seattle residents. But he, the city council, and the staff at the Utility all seemed to recognize that if residents were to be successful in their new roles — as sorters, recyclers, and conscious consumers — then city government services would also have to change. In other words, this was not simply the individualization of a collective problem, and it did not, as might be predicted, absolve the local government of its responsibility. Rather, Seattle committed itself to providing the infrastructure and programs necessary to allow resident recycling to succeed. So while the emphasis on recycling was driven largely by resident demand, the city leaders agreed to invest substantially in outreach, in education, and in the development of programs that would be effective, convenient, and robust. The city would have to provide visible, legible, convenient recycling service, and it would have to train Seattle residents in these new systems.

Some in city government pushed back against the arguments for voluntary recycling. The city council's Environmental Management Committee, responding to calls from ambitious residents, was more aggressive. Its members argued that, given the city's bold vision, recycling should be mandatory. Perhaps concerned about the enforcement of a mandatory program, or about the degree of change already occurring in the system, Diana Gale told the city council that she preferred voluntary programs. Seattle citizens were enthusiastic about recycling, she said, and "we think they should have the chance to voluntarily become the world leaders in recycling without behavior change being legislated upon them."[84] The mayor took this argument further, contending specifically, in relation to yard waste separation, that "forcing mandatory ... separation on citizens before they have had a chance to respond may alienate some customers. Requiring separation without budgeting for an enforcement program will make them cynical."[85]

The city council staff ultimately concurred with Gale and the mayor and confirmed a voluntary recycling approach in 1987.[86] As a contingency, the resolution on the subject allowed for a move to mandatory recycling, should it prove necessary. Ultimately, the city did move to mandatory yard waste separation, then to mandatory recycling, and eventually, in 2016, to mandatory food scrap separation. But throughout the debate about voluntary recycling, citizens, advocacy groups, and many city government officials aligned around newly conceived roles and responsibilities for both individual and state actors in the waste system.

In 1988, the Utility, city council, and mayor uniformly agreed on a course of action: to move ahead independently from the county and develop a recycling-dominant system with no waste incineration. To make this plan work, the Utility contracted with a newly built, low-cost landfill in Eastern Oregon. And here again, Seattle's resistance to dominant trends, and the depth of its redefinition of waste, was evident.

The city wrote a savvy contract that allowed for waste reduction over time. Even though the city was contracting with a private corporation for disposal for the first time, it structured the contract so as to maintain control over costs and incentives. The initial tonnage rate was less expensive than an incinerator would have been, and the contract specified that there would be no penalty for reduced tonnage. As Tim Croll, a Utility project manager at the time, described the deal, the contract said, basically, "we are willing to take this stuff, put it on a rail car, a train, and take it [to Eastern Oregon] and bury it, landfill it. But we will not demand any put-or-pay. If you have a magic recycling machine, we'll just take the [remaining] 10 percent. You won't owe us anything for any missing tonnage."[87] This kind of contracting was unusual at the time and indicated the degree to which waste managers had reinterpreted their responsibilities. Even though Seattle's system still operated comfortably in the space established by the WRWR—i.e., focused on end-of-pipe, imposing costs only on individuals and governments — the landfill contract institutionalized a value set that made space for meaningful changes in material flows over time.

To mark the decision, the mayor wrote a letter to the city council making clear that he understood the significance of Seattle's "exciting and aggressive" plan:

> We have committed the City to provide convenient programs to accommodate the citizens' resolve to recycle; we have chosen the use of an environmentally sensitive and cost effective landfill for disposal of nonrecycled waste; and we have exercised our independence from King County to alleviate the burden on County disposal facilities and to provide rate stability for both residential and commercial customers. Never in recent times has Seattle enjoyed such a broad consensus on handling the City's trash.[88]

Through its extensive and patient public dialogue, Seattle managed to arrive at an unconventional but highly popular course of action. In so doing, and over the course of just a few years, the city's response to the disposal crisis had turned almost completely inside out.

Over the course of the 1980s, the problem of waste in Seattle migrated from a straightforward disposal problem to a complex spatial, political, temporal, and material resource problem. Public officials, at the urging of citizens, began to understand garbage as a valuable material containing a variety of potential resources and to recognize its management as a source of political power. As the definition of garbage and the garbage problem shifted, so did the understanding of who was responsible for the solution and what suitable solutions would look like. Responsibility for solving the problem shifted, moving from individuals and the city government — each acting alone in roles prescribed by the waste regime — to a robust partnership between them.

The process of problem redefinition occurred through both formal planning processes and informal dialogue among state and citizen actors. Through this process, Seattle abandoned an incinerator plan in favor of aggressive recycling and waste reduction — defying the dominant logic of privatization and techno-managerialism that characterized both solid waste practice and many other arenas of public administration at the time. Seattle's new system asked more from its citizens, echoing in some ways the deflective and cynical messaging of packaging industries. But individual action is not in itself problematic — individuals must be responsible for their consumption. It is only problematic when calls for individual responsibility do not recognize the structural barriers to behavior change. One piece of this puzzle is, of course, the fact that production remains largely unregulated. But another piece is the existence of convenient and functional back-end recycling infrastructure. In Seattle, increased citizen engagement in waste management was accompanied by increased demands on disposal and recycling contractors to ensure that recycling activities were possible and convenient and that materials collected were actually recycled. The city never abdicated its own role or responsibilities; rather, it reframed and expanded them to support new modes of materials management.

While the waste industry was organizing to make waste more invisible, more distant, and more profitable, and manufacturers and retailers were developing and marketing ever-growing quantities of increasingly disposable consumer goods, Seattle pushed back. Seattle's waste plan demanded that individuals become more engaged, not less, with the processes of wasting. With its decision to prioritize recycling, Seattle defied the WRWR and set the stage for even more radical action that would eventually implicate producers in the network of actors responsible for the "garbage problem."

CHAPTER 4

Compliant and Defiant Wasteways: Boston and Seattle within the WRWR

At the beginning of the 1980s, Seattle and Boston were in a similar predicament, but by the end of the decade they were in totally different places. Seattle had produced a new way of defining and managing waste, a defiant wasteway that chafed against the weak recycling waste regime. Boston, meanwhile, had constructed a compliant wasteway, and remained tethered to the WRWR. A city's wasteway—the relationship between its waste system and the regime in which it is situated—is shaped by how people and waste management organizations define waste; by who participates in that act of defining waste and waste problems; and by the roles and responsibilities of system managers and regular citizens.

The compliant wasteway that emerged in Boston through its waste planning process operated more or less as one might expect a creature of the weak recycling waste regime to function. Characterized by limited and incremental changes driven by actors with the most power in the system, it had the hallmarks of a modernist socio-technical regime. Garbage was defined as pollution, dirt to be removed. Until 2018, when the city launched a potentially transformative waste planning initiative, changes in the city's waste system did not challenge the fundamental relationships between citizens, the state, and private-sector service providers, and certainly did not challenge the extraction-manufacturing-consumption-waste chain. Beyond contributing the waste to be managed by the system, actors outside of city government and industry had only a limited role in waste planning or decision-making.

The key transformations that took place in Seattle, in contrast, created a defiant wasteway that challenged key fundamentals of the WRWR. The defiant wasteway positioned Seattle as a national waste management leader. Perhaps more significantly, it launched the city into a condition of constant evolution, incrementally moving in a different direction from the waste regime within which it was embedded. The redefinition of waste in Seattle led to the creation of a new set of institutions for governing waste. These institutions incentivized recycling, reuse, and reduction through service and infrastructure investment. In turn, citizen engagement with waste planning and management interfered with cultural and market-driven narratives of convenience and also "de-othered" garbage—giving it positive material qualities. Seattle's emphasis on material life cycles challenged the indomitable one-way march of disposability and planned obsolescence.

The waste system that evolved in Seattle was highly experimental and often encountered failures. When it pushed too far, it was disciplined, either by industry or by citizens themselves. But because the city's waste system transformation relied on solid institutions and a deep base of creative professional support, it has proven resilient. As a landscape in succession, the evolution of the defiant wasteway in Seattle has led to a system that is increasingly different from the waste regime within which it operates.

This chapter first traces the arc of Boston's compliant wasteway through a series of system changes that did little to alter the social, economic, or material profile of waste management in the city. It then follows Seattle's defiant wasteway through increasingly radical changes to the city's system, including some failed experiments, that nevertheless have added up to a waste system that resists the WRWR.

Boston's Compliant Wasteway

Boston's waste system did not remain static after the final decision to export waste to an incinerator in the suburb of Saugus, Massachusetts, but it remained fixed within a compliant wasteway. Subsequent programmatic changes in the city only reaffirmed the waste regime. The city did not revisit system goals, nor did it engage in any substantial public discourse about waste management, until 2018. The Public Works Department (PWD) continued to lack capacity for planning or inclusive decision-making, remaining reliant on industry partners and waste engineers for leadership, innovation, and day-to-day operations. The changes that did occur echo research findings about innovation within socio-technical regimes: change over time was incremental and reinforced existing power structures and interests.[1] Boston's introduction of curbside recycling in the early 1990s, and the subsequent transition from dual-stream to single-stream recycling, demonstrate how

Boston's compliant wasteway reproduced the weak recycling waste regime on the ground.

After Boston opted for garbage export in the late 1980s, some city officials hoped that the South Bay incinerator proposal would be replaced with a state-funded, high-capacity, modern recycling facility. In fact, on the heels of the incinerator decision, Mayor Raymond Flynn announced that the city would soon start a recycling program, in part because the state had committed to building a material recovery facility (MRF) in Boston alongside the Suffolk County House of Correction. The city did not expend resources on a recycling plan, and the Dukakis administration did end up constructing such a facility elsewhere in the state, but eventually declined to build one in Boston. The city thus initiated its first formal recycling program on its own, without state support, without centralized recycling infrastructure, without a developed market, and without a comprehensive strategy.

When Mayor Flynn finally agreed to introduce a city-run curbside recycling program in 1988, it was mostly motivated by the promise of jobs. A paper mill in Hyde Park closed in the early 1980s, throwing 135 people out of work. When a recycled paper manufacturer from California made an offer to buy the mill and reopen it, Boston's public officials suddenly got serious about recycling.[2] But city officials' interest in recycling was couched in the same caveats as always. Flynn reiterated to reporters that he believed "the primary trash issue in Boston is not the particulars of how to handle the city's waste that can be recycled . . . but forcing the state to come up with a solution for the waste that cannot be recycled."[3] He had resisted recycling on this logic for the better part of a decade. While supportive of recycling in theory, the Flynn administration had always argued that recycling was a "percentage game": regardless of how much material was recycled, there would always be a substantial portion of waste requiring disposal. To emphasize recycling, therefore, would be irresponsible.[4] This is the logic of the weak recycling waste regime, and it was on full display in both the recycling ordinance that the city eventually passed and in the rollout of the city's first curbside recycling program.

Flynn finally signed a recycling ordinance into law in July 1990, and it included some relatively ambitious recycling goals: 28 percent of the city's residential waste stream, by weight, by 1991, 38 percent by 1995, and 50 percent by 1998. There had been substantial ambivalence about these goals—the mayor had wanted to reduce or eliminate fixed targets altogether, but local environmentalists pushed hard, arguing that they would help to increase the city government's commitment to implementation.[5]

In a compromise, the ordinance included a provision that if there was insufficient capacity at regional MRFs (another dig at the state for its failure on regional waste infrastructure) for collected materials, then lower recycling targets would apply. The ordinance mandated public outreach to inform residents about the curbside program, but it left all the details of program design and implementation to the commissioner of public works, Joe Casazza. Recycling was voluntary, although the ordinance did permit haulers to stop collecting recyclables from specific participating households if, after two written warnings, nonrecyclable materials were found in their curbside containers.[6]

After the passage of the ordinance, Flynn established a Recycling Advisory Committee, whose composition was similar to that of the Citizen Advisory Committee on Solid Waste Disposal a few years prior—business and neighborhood representatives. In a departure from the norm, the committee also included experts on recycling and the environment. One such expert was Sue Minter of the Boston Recycling Coalition (BRC), an organization made up of environmental and labor interests advocating for recycling and waste reduction as primary solutions for municipal waste management. The BRC had run a volunteer recycling program in the city for years. The Recycling Advisory Committee also included Amy Goldsmith, from the environmental advocacy organization Clean Water Action (CWA). CWA was part of the Boston Recycling Coalition and a strong advocate for progressive urban environmental policy. Both Minter and Goldsmith were pleased with the city's recycling ordinance, viewing it as a step in the right direction after a decade of stalling. Goldsmith believed that recycling could reduce the need for any new incinerators in Southeastern Massachusetts. Minter, for her part, hoped that residents would be relieved that they no longer had to store and haul their own recyclables to monthly drop-off sites.[7]

The presence of recycling experts on the task force could have been transformative, but Flynn's WRWR-inflected priorities circumscribed the program rollout. He was committed to a recycling program that would be both "environmentally strong and fiscally prudent." In exchange for accepting the recycling targets proposed by environmentalists, Flynn obtained a provision in the law mandating that recycling not increase the net costs of solid waste management for the city.[8] Whereas Seattle had adjusted its cost-benefit calculations via the true cost and avoided cost frameworks, Boston instead committed to never paying more than current disposal rates for recycling. The fact that Flynn wanted a fiscally prudent program is not in itself unusual. But the fact that the existence of the recycling program was tied directly to disposal costs shows the extent to which disposal governed the recycling program. The recycling ordinance was structured by the priorities of the WRWR; it was a product of a compliant wasteway.

The WRWR was also evident in the limited curbside recycling program that was eventually rolled out. The city did not have to start from zero. As explored in chapter 2, there were many people and organizations actively promoting recycling as an alternative to incineration. The city even received economic development proposals from coalitions of businesses and nonprofits seeking to help the city develop markets for recycled materials. And, over the late 1980s, the BRC had run a network of volunteer-led drop-off recycling centers across the city. BRC volunteers ran monthly and daily drop-offs at five locations. Thousands of city residents brought their recyclable discards to them; the PWD even assisted by renting containers and contracting with haulers to collect the sorted materials from the drop-off centers. Three types of plastics were collected and sorted onsite, as well as newspaper, cardboard, glass, and aluminum.[9] The BRC had done all the legwork to find buyers for the materials, had educated residents, and had started to build a real market. But once it initiated its own program, the city abandoned this foundation.

On September 16, 1991, Boston launched the city's first curbside recycling program. It began as a small pilot in one part of the Jamaica Plain neighborhood and eventually was expanded to residential buildings citywide. When the pilot program launched, the PWD ended its assistance to the drop-off centers (which, it made a point of noting, had cost the city a total of $3,000 over the course of several years, but had led to $58,000 in saved disposal costs and $61,000 in saved hauling fees) and replaced it with every-other-week collection of newspapers, shrinking both the volume and types of material that could be recycled in the city.[10]

In the weeks leading up to the curbside program, city officials tried to spread the word. The PWD engaged college student volunteers, a local non-profit, community organizations, and the well-connected members of the Recycling Advisory Committee to distribute flyers. The city ran stories and ads in local papers and distributed blue recycling bins to every household in the neighborhood where the program launched.[11] Despite these efforts, many residents remained uninformed about the new program. The launch of the program was so quiet that one loyal recycler, Dorchester resident Margaret Toro, wrote to the mayor to say that although residents welcomed the program, it had "been designed to fail." She cited two reasons for her assessment: first, most residents had received no information about the new program; second, the initial biweekly schedule was inconvenient and difficult to remember, particularly because the city only provided the date of the first pickup and no schedule for ensuing collections. Mayor Flynn forwarded the letter to Joe Casazza, who replied to Toro that the PWD was doing the best it could with a "very limited budget" for outreach. He enclosed a new flyer with the collection schedule.[12]

Toro's assessment highlights how the new program in Boston contrasted with similar program launches in Seattle. First, the lack of resources for communicating with residents demonstrated that the city was not prepared to invest in citizen outreach or participation, a signal that it sought to maintain preexisting structural relationships. Indeed, the quiet recycling launch showed that the city's waste managers viewed citizens not as a significant part of the waste system, but instead as waste generators and service recipients, the roles prescribed by the WRWR. The bottom line was that recycling was a subordinate part of waste management, carrying forward the implicit problem definitions and end-of-pipe orientations of the WRWR. Boston's compliant wasteway reinforced and reproduced those definitions, enacting the abstractions of the WRWR in urban space.

Other pieces of the curbside program further reinforced the traditional organization and priorities of the city's waste system. For instance, at the recommendation of Joe Casazza, the PWD amended the contract with its waste hauler, BFI—one of the country's largest waste management firms— to include recycling collection, rather than finding a contractor with recycling expertise.[13] Making a minor contract amendment in lieu of investing in the right expertise is a typical product of WRWR thinking. Casazza viewed the recycling program as a minor amendment to garbage hauling, just an additional program in service of waste disposal.

A *Boston Herald* article in 1991 reported on the results of a Massachusetts Public Interest Group (MassPIRG) survey, which found that Boston not only had among the lowest recycling rates of large cities in the United States, but also spent the lowest amount per capita on recycling programs and employed the fewest staff. In response to criticism about the disappointing rollout of the city's recycling program, and in continuation of the pattern of politicking from the South Bay incinerator debate, the Flynn administration passed the buck to the state. The article noted that a Flynn administration official said that "Boston would now be recycling 25 percent if Flynn's proposal to build a Materials Recovery Facility at South Bay had not been vetoed in 1988 by the Dukakis administration." This comment must be understood as pure politics. The Flynn administration's plan was never more substantial than a few words to the press. According to MassPIRG, the lack of recycling was due almost exclusively to underinvestment by the city. Boston residents, MassPIRG noted, had already shown an "unusual willingness to participate in recycling programs." The *Herald* even compared the dismal recycling rate in Boston, then at 7 percent, with that of Seattle, which by 1991 was diverting 44 percent of residential solid waste from the landfill.[14]

Residents from neighborhoods excluded from the initial curbside recycling pilot program wrote to Commissioner Casazza and the city council about their individual organizing efforts and requested that the city's recycling program be expanded.[15] Upon recognizing the widespread support for recycling among residents (which city officials might have understood at any point in the preceding decade, had they been listening), the city did increase the budget for the program during its second year, despite widespread budget cuts in other areas across the board. The recycling program was extended citywide, and residents could eventually recycle paper, glass, plastic, and aluminum curbside. The city's investment, however, did not alter the city's wasteway. The same professionals remained at the helm. The same contractors provided services. There was no strategic planning, no wider vision. The goals established in the ordinance were never met, and the city never reported on its progress or revisited the initial targets. The program remained narrowly defined and implemented, functioning just as a WRWR recycling program should.

Boston residents' long-standing interest in more aggressive and transformative waste management never dissipated. After the disappointing initiation of curbside recycling in the early 1990s, various state, nongovernmental organization, and industry actors worked at different levels to amend the city's disposal-oriented system. Only the private sector, however, succeeded in instituting substantial programmatic change. In 2009, Boston shifted from dual-stream to single-stream recycling.[16] Both the outcomes of the shift and the way it unfolded reinforced the WRWR.[17]

The shift to single-stream recycling meant that residents could comingle all recyclable paper, plastic, glass, and metal in a single wheeled cart, rather than separating paper and cardboard from the other materials. Single-stream recycling is a disposal-industry-driven WRWR technique. Waste companies that provide single-stream service must invest in expensive, high-tech MRFs that are able to separate materials with minimal human labor. In this sense, single-stream recycling is like incineration: it is favored by the disposal industry because it is technologically advanced and capital intensive. Only the large companies can provide the service, so this option helps them to consolidate market share. But single-stream programs serve the production end of the regime as well: the Recycling Partnership and similar organizations advocate for single-stream programs because this approach makes recycling easier for the consumer. It further alienates people from their discards. Single-stream recycling maximizes convenience and allows people to imagine that absolutely everything is recyclable, so that they don't have to worry about what they purchase in the first place. Single-stream recycling also increases contamination, reducing the total percentage of collected material that is actually reprocessed and converted into new consumer products. But since material conservation is not the objective of either the producer or the disposer coalitions in the WRWR, these problems have not stopped either from aggressively pushing the single-stream approach as a recycling solution, even going as far as to offer grant support to communities to make the switch.[18]

Boston made the shift to single-stream recycling at the same time as several other cities in the region. The private firm that held the recycling collection contract for residential waste in the city had retrofitted its sorting plant to accept comingled materials and encouraged its customers to make the switch. Boston had no budget for programmatic innovation but was finally enticed to shift to single-stream recycling in 2008 when a private company offered a free trial of sixty-four-gallon recycling carts. With the trial carts, the city implemented a single-stream pilot in the Jamaica Plain neighborhood. Residents in other neighborhoods started requesting the larger, more convenient carts almost immediately. But what really impressed the city's waste managers "was ... the cleanliness of it." The PWD's sanitation superintendent "really liked the cleanliness part and wanted to take the carts that we had left over to a downtown neighborhood that has a real trash problem."[19] The tidiness of the new, larger, single-stream carts was a critical factor in convincing waste managers to find the funding to take the program citywide. In a holdover from the Sanitary City waste policy, cleanliness and efficient removal of waste continued to characterize Boston's system even after Commissioner Casazza's tenure.

The new single-stream program did lead to an immediate 45 percent increase in the amount of recyclable material collected in Boston, a leap that city officials viewed as an indicator of success.[20] The increase, though, only took the residential sector from a 12 percent to a 17 percent recycling rate, still well below the national average.[21] The service change did not affect the commercial sector, and it did not account for increased contamination. The shift to single-stream recycling was not about changing material flows through the economy—even though advocates, and the city's under-resourced recycling director, may have hoped for that. Boston's waste managers continued to view their primary role as a protector of urban hygiene, and they continued to justify decisions that served producer-disposer interests in those terms. Single-stream recycling in Boston showed, in other words, how the definitions, processes, and interests that dominated the incinerator decision continued to influence waste policy for decades, keeping the system locked within the WRWR.

Seattle's Defiant Wasteway

Scholars of socio-technical systems have long recognized that large, technical systems are difficult to change. Sustainability scholar Gregory Unruh has argued that, in locked-in technological systems, institutions, industries, and technologies "interact to create a self-referential system that tends to increase in value with the growth of the technological system."[22] Historian Thomas Hughes argued that this kind of self-reinforcing feedback "resist[s] changes in the direction of development," a phenomenon he referred to as "technological momentum."[23] Compliant wasteways, like systems described by technological momentum or lock-in, evolve along predictable pathways reinforced by powerful interests. The case of Seattle indicates that "momentum" can also characterize a self-reinforcing system that moves in an opposing direction. The city's defiant wasteway has continuously adapted to new conditions and responded to external stimuli; it has recalibrated when disciplined by the waste regime, and it has moved incrementally toward ever more ambitious targets, all the while strengthening institutions that resist a return to a more conventional system.

Seattle's defiant wasteway relentlessly advanced its targets for waste reduction and diversion even when its goals proved elusive, all the while innovating programmatically. In tracing the broad contours of system change after the 1980s, we can see how the city experimented, adapted, and reacted when disciplined. Over the late 1980s and through the 1990s, it institutionalized its wasteway through a series of new ordinances and resolutions that established key frameworks for monitoring, planning, and goal setting. After Seattle adopted Resolution 27871—which established a recycling goal of 60 percent by 1998—and officially rejected WTE, the Solid Waste Utility, the mayor, the city council, Seattle residents, Seattle businesses, King County, the State of Washington, and a variety of allied and contrary interests continued to negotiate, experiment, plan, and evaluate, all while slowly making progress to turn away from the weak recycling waste regime and toward Seattle's ever more ambitious goals.

In 1998, city councillors affirmed the 60 percent diversion goal, but pushed the date to 2010; in 2004, they affirmed the 60 percent goal again, but pushed the date to 2015. Despite the difficulty of achieving 60 percent, the city continued to push for more ambitious heights. Resolution 30990, which the city refers to as the "Zero Waste Resolution," adopted in 2007, tweaked the diversion goal again, seeking a 60 percent diversion by 2012 and a 70 percent diversion by 2025. It also said Seattle would not dispose of any more solid waste than the amount that went to the landfill in 2006, establishing the city's first cap on solid waste disposal. The resolution required the Utility to produce an annual report on the city's progress toward these goals as well as its short-term plans for coming years.[24] Four years later, the city council adopted a new comprehensive plan for solid waste (titled "Picking Up the Pace to Zero Waste [2011 Revision]") that amended diversion targets even further: it required recycling 60 percent of the city's solid waste by the year 2015 and 70 percent by 2022. For the first time, Seattle also established a recycling goal of 70 percent by 2020 for construction and demolition debris.[25]

In the summer of 2016, in a report released exactly on schedule, Seattle Public Utilities reported that the city had recycled 58 percent of its commercial and residential waste in 2015 — it had gotten achingly close to the 60 percent goal, but was not quite there.[26] Over the course of the past three decades, the city has consistently fallen short of its ambitious recycling goals. That the city constantly recalibrates the target years, however, suggests that the gap between goal and reality is not representative of failure, but evidence of determination. By tugging the 70 percent target year ever sooner, the city has recommitted to striving toward the goal even while adding percentage points has become increasingly difficult. And, while recycling goals remained elusive, the city reduced waste generation nearly 13 percent between 2006 and 2018 despite a growing population. Overall, the amount of Seattle's waste being landfilled dropped more than 30 percent between 2006 and 2018,[27] and that number is more meaningful environmentally than the recycling rate alone.

Seattle's 1998 solid waste plan, titled "On the Path to Sustainability," provided a detailed roadmap to Zero Waste, making Seattle among the first cities in the world to formally declare Zero Waste as a realistic municipal goal.[28] Each subsequent plan and plan revision has advanced Seattle's ambitions even further. The 2013 plan update recognized solid waste as a complex socioeconomic phenomenon that is shaped by programs, physical infrastructure, and attitudes. It presented a clear vision that engaged all aspects and actors in the system. The plan addressed the impacts of the 2008 recession and recovery on the waste stream as well as the implications of a mostly high-end, multifamily housing boom. It included an entire chapter on waste prevention that looked equally to behavior change, city and state policy frameworks, and infrastructure development to advance the tough project of generating less garbage at all levels.[29]

In addition to an institutionalized system of ambitious and evolving targets, Seattle used monitoring and reporting as tools for public accountability. Many cities, including Boston, have formally adopted diversion rate goals that are meaningless, because no one ever assesses how close the city is to achieving them. In Seattle, the waste reduction and diversion goals are relevant to everyday waste management and policy because of consistent attention to and investment in monitoring, reporting, and contingency planning. Monitoring and reporting play a key role in the defiant wasteway; without them, the Seattle public and its elected officials would have no way of gauging and celebrating progress, or knowing where to invest attention and resources. In addition to requiring monitoring and reporting through resolutions and ordinances, city council resolutions also frequently ask for contingency plans in the event that goals aren't met. The combination of monitoring, reporting, and contingency planning reinforces the city's momentum and thus plays a key role in constituting the city's defiant wasteway.

A September 1987 resolution required the Solid Waste Utility to produce an annual recycling report that would establish interim annual recycling goals toward the larger goal of 40 percent diversion by 1997 and "provide contingency plans if those goals are not met."[30] Furthermore, it directed the Utility to develop a database and monitoring mechanism to assess recycling program performance. The emphasis on monitoring, reporting, and contingency planning—which could also be interpreted as technocratic governing tools—at this stage actually indicated an understanding that the city's ambitious goals would not be achieved quickly or abruptly. Rather, the nascent institutions of the new wasteway were being structured to accommodate progress and regress, changes in context, new directions, and as yet unimagined possibilities. When the city council adopted a 60 percent diversion goal a few years later, it maintained the emphasis on monitoring and contingency planning. The resolution established that the recovered tonnage "shall be monitored at regular intervals to check on the City's progress in meeting its goal." Then, if particular programs were not on track to meet goals, "the Solid Waste Utility shall recommend a new or revised program that will accomplish the City's 60 percent goal."[31]

The processes of monitoring and contingency planning allowed for political accountability. They provided a way to daylight the traditionally black-box operations of waste management. They also provided a way to involve Seattle's residents in the achievement of the city's goals, reinforcing that these were collective ambitions and not just the purview of technocratic insiders. Each year, the Utility's recycling reports were released with press events.[32] The reports are available—back to 2007, when the annual recycling reporting requirement was formalized—on the Utility's website. These reports, along with a series of other regular reports on operations and recycling market conditions, form the backbone of program development.

As services expanded, Seattle residents became accustomed to a more intimate relationship with their discards. Norms around waste and waste handling shifted as a product of ongoing public conversations about garbage. Norm shifts are evident in everyday behaviors of Seattle residents and in the evolving expectations about system performance. Residents in single-family homes routinely recycle or compost more than 70 percent of their household trash, a feat that requires thoughtful treatment of everything from coffee grounds to worn-out boots to pet carcasses.[33]

While average household waste-handling practices in Seattle are quite different today from what they were in 1990, the Utility has rolled out changes slowly, which resulted in the normalization of shifts that might have appeared extreme or radical at other times or in other contexts. In fact, when Seattle residents travel, they sometimes experience a distinct kind of unease at the idea of throwing perfectly compostable food scraps into a landfill-bound trash can. Utility managers have heard feedback from Seattle residents who are dismayed by the lack of recycling and composting infrastructure in other cities. City leaders count it as a success that for many Seattle residents, throwing organic material into a garbage bin now simply feels wrong. It is a misplacement, a disordering, like littering.[34] As new rules have unfurled, residents have, by and large, gamely complied.

To be clear, although residents of Seattle were always good recyclers, the shifts in household practices cannot be attributed to some preternatural affinity for sustainable waste handling. One local newspaper surveyed its own trash in 2015 to discover a rather embarrassing quantity of recyclable material in the garbage cans.[35] And there are plenty of residents who have complained over the years that the city's tactics were heavy handed. Don Hannula, a regular columnist for the *Seattle Times* in the 1980s and 1990s, voiced such an opinion for decades, with snarky commentary on Seattle's "despotic" waste management techniques.[36] But the incremental introduction of new practices, along with financial incentives, carefully considered education campaigns, and, eventually, legal requirements with reliable enforcement, swept most residents along in the creation of new habits.

As residents acquired new habits, took on new relationships with their discards, and began to abide by new laws, Seattle's waste managers also continued to redefine the role of the state and to reshape citizen expectations for state participation. In 1988, the city council passed Resolution 27828, which directed the city government to procure recycled products, to reduce the use of single-use items, and to recycle. As part of the larger redefinition of the roles and responsibilities of citizens and the state, this resolution described the city government as a consumer and waste generator (in essence, a citizen) with the same responsibilities as citizens themselves. In a direct articulation of this role, the resolution stated that the "Seattle City Government has a responsibility as a member of the greater community to develop policies and practices for its own operations to assist the City in achieving its recycling and composting goals."[37]

Resolution 27828 was followed by another resolution formally adopting specific policies recommended by the mayor to achieve Resolution 27828's goals. The recommendations included procurement of recycled paper and recycled paper products, a plan to phase out Styrofoam and plastic beverage containers (shifts in consumption!), department-by-department waste reduction plans, and increased recycling in city departments.[38] These resolutions were soon followed by an ordinance banning city purchase of polystyrene foam and beverages in plastic containers.[39]

Through these resolutions and ordinances, the city imposed new limits on government action. As part of the effort to make the city government into a better citizen, the Utility developed educational material to train city employees across departments. One such pamphlet explains the city's goals and programs and offers tips on how city employees can recycle both at work and at home. The pamphlet, titled "City Employees Recycle," addressed municipal workers as employees *and* citizens.[40] The flyer shows how Seattle's waste managers sought to maximize the impact of their educational reach and worked to establish broad-based waste habits across different urban spheres. The city government was reckoning with its role as a consumer, coordinating among departments and agencies that typically operated independently, acknowledging its employees as citizens, and participating fully and actively in the pursuit of citywide waste management goals through multiple channels outside of its traditional role as service provider to residents.

As the roles and responsibilities of both residents and the city government evolved, the city moved away from the reliance on voluntary action that characterized the first generation of the defiant wasteway. Both the mayor and the Utility remained committed to voluntary action throughout the 1980s, even while some city councillors pushed for mandatory recycling. The mayor and Utility leaders hoped that with convenient service and educational outreach, Seattleites would be compelled to recycle. This was true to a point; the city saw a spike in recycling rates after curbside recycling was extended to the entire city. However, after the initial spike, participation rates plateaued. Waste managers began to consider new methods to increase recycling, including exerting more control over commercial waste and making recycling mandatory. Changes were slow in coming, and they were thoroughly studied. In 1992, the city council adopted a resolution that permitted the Solid Waste Utility to develop recommendations for exerting local regulatory control over commercial waste, which, according to the resolution, constituted 60 percent of the city's waste stream. The resolution established six potential options, ranging from no change to complete takeover by the public sector, and instructed the Utility to study them all.[41] The Utility did the studies but did not make any immediate service changes.

The city finally instituted mandatory commercial and residential recycling in 2005. Mandatory recycling was highly controversial during the 1980s, and interfering with the commercial sector remained controversial through the 1990s. But they were both an easy sell in the early 2000s. Seattle's mayor at the time, Greg Nickels, proposed new mandatory recycling rules to the city council in 2003, and the council passed them unanimously, with little debate. Mandatory recycling of paper and cardboard for the commercial and residential sectors took effect on January 1, 2005.[42]

The development of enforcement tactics was a key issue in the transition to mandatory recycling, particularly as the city was attracting many newcomers as a result of the booming tech industry. The Utility was thoughtful about how to get residents old and new up to speed. Over the first year, residents received warnings and notifications about the new rules. Starting in 2006, residents with too much paper and cardboard in their trash received three warnings and then a $50 fine. The Utility was careful about how it conducted inspections. Bret Stav, a spokesman for the Utility at the time, told the *Seattle Times*, "We're not training people to go in with a magnifying glass. Inspections will be based on immediate visual recognition."[43] While some residents struggled to adapt to the new rules, Tim Croll, then solid waste director for Seattle Public Utilities, told reporters that the Utility had received lots of questions about how to comply, but very few complaints.[44]

The city rolled out new organic waste collection containers that could accommodate both yard waste and food scraps in 2005. Not all residents were entirely sanguine about having to separate food scraps in addition to recycling. To ease the transition, the city began with a voluntary citywide program. The Utility distributed bins, conducted a widespread public information campaign through local media, and provided educational materials directly to households.[45] There was an immediate 44 percent increase in the

Figure 4.1. The trash cans for a Seattle household subscribed to the micro-can rate. Every household receives the same recycling and organics bin but chooses the size of its trash bin (the small black box pictured here). There are larger can options available. Photo by author, 2014.

curbside collection of organics in 2005.[46] In 2014, the city council unanimously approved mandatory composting; the new rule took effect in January 2015 with little fanfare.[47]

Each of these incremental steps—the distribution of new containers, the educational campaigns, and eventually, the establishment of mandatory participation, along with enforcement procedures to ensure compliance—was carefully calibrated to nudge the existing system away from conventional waste management and toward ambitious goals. With each incremental step, Seattle residents grew used to new processes. Elected officials endured hiccups along the way, and the Utility responded to problems and complaints. In each case, after an initial flurry of press, the changes faded into normalcy.

While Seattle slowly rebuilt its regulatory framework, implemented new programs, and constructed a new social context for waste-making and waste management, it also redesigned the physical infrastructures of waste collection, sorting, and treatment. At the smallest, most distributed scale, Seattle provides uniform garbage pails to all single-family residences (see figure 4.1). Since the late 1980s, when residents began to choose their level of garbage service (how much garbage they would pay to throw away every week), they have received garbage, recycling, and organics bins from the Solid Waste Utility.[48] With these bins, people can easily see how wasteful they are compared to their neighbors, which amounts to a form of peer pressure. Research has consistently found this approach to be effective for encouraging recycling participation.[49] The uniform bins also provide an additional layer of connectivity between the Utility and its customers.

The relative sizes and costs of carts are fully integrated into the city's system goals. The cost of the different subscriptions is supported by the size of the different cans, and the extra cost for the privilege of throwing away more waste serves to support the goal of getting people to produce less garbage. In a signal that the pricing is effective, 14 percent of households in 2019 subscribed to the micro-can rate; 30 percent to the mini-can rate, 51 percent to a single can, and only 5 percent to a two-can rate, putting Seattleites well ahead of others in waste minimization.[50]

Over the past decade, Seattle has rebuilt its two city-owned transfer stations for the same purpose. Like the city-provided bins and carts, the transfer stations are the physical presence of the Solid Waste Utility in the city, and as such they are materially and symbolically significant. Originally built in the 1960s, the two stations were designed to consolidate the city's garbage for efficient transfer and disposal. The stations were huge sheds built around enormous pits into which garbage was dumped and then compacted. This design made material separation next to impossible.[51]

The city's 1998 solid waste plan, "On the Path to Sustainability," identified a series of inadequacies at the transfer stations. It argued that the facilities lacked adequate capacity; were outdated, smelly, and far from earthquake ready; and were unable to handle a changing waste stream. The Utility launched an intensive planning process that yielded the 2003 Draft Solid Waste Facilities Master Plan, which was ultimately folded into the city's 2004 amendment of the 1998 "On the Path to Sustainability" plan. Both the planning process and the design outcomes showcased Seattle's defiant wasteway.

The planning process for the transfer stations was characteristically intensive and open. It included meetings and other avenues for input specifically targeted to neighborhoods, city residents, and Utility customers; advocacy and civil society organizations; transfer station employees; King County officials and staff; waste haulers; businesses that used the transfer stations; and solid waste, recycling, Zero Waste, and other industry experts from across the country.[52] Once they were established, the goals were then codified by city council resolution.[53] The breadth of outreach was characteristic of the Seattle Public Utility's planning approach.[54] As with the critical planning processes throughout the 1980s, this process seriously engaged expertise from a variety of sources, including waste workers themselves, who have traditionally been marginalized in planning and policy.[55] Recommendations from these sectors ran the gamut from concern about worker safety to the logistics of temporarily closing a transfer station and how to effectively design for better materials management.

The goals for the transfer station reinforced and advanced many of the core shifts that took place during the 1980s, including an emphasis on balanced fiscal and environmental costs and a recognition of the specific temporal, spatial, and material characteristics of waste. The plans proposed that the transfer stations' physical design be consistent with goals and priorities established through the city's regularly updated comprehensive solid waste plans and other city plans. The 2004 comprehensive plan amendment, specifically, was a key agent in promoting more ambitious goals. Pegging the facilities plan to it ensured that the physical infrastructure would support the city's broader vision for materials management.

Once completed, the transfer stations embodied the ongoing evolution of the waste system. In 2013, Seattle completed the renovation of the South Transfer Station near South Park, a $75 million project that rebuilt the site from the ground up. From the outside, you could be forgiven for mistaking the facility for an arts center or a building for some other kind of creative or civic use. The sidewalk to the entrance is bordered by a wall with nooks displaying works of art created by Seattle-based artists from waste materials. The wall of the main entrance blazes green from decommissioned street signs (figure 4.2).

The station is a public-facing facility designed to make material separation easy and natural. Though it is more enclosed than its predecessor, in order to control odor, it is also more transparent. The facility contains a viewing gallery, and members of the public can enter at any time to watch the dumping and sorting in progress on the facility floor. On the glass-enclosed overlook, text bubbles provide visitors with basic statistics about waste collection in the city and the amount and type of material that moves through the transfer station. The gallery contains exhibits about the city's waste system, about the science of composting, and about new waste diversion and prevention programs (see figure 4.3). In one exhibit, called "Rethink—Trash or Treasure," panels with photographs of common waste materials open to reveal facts and describe alternatives to traditional waste disposal. A photo of a disposable plastic water bottle, for instance, opens to

Figure 4.2. The main entrance of Seattle's South Transfer Station. The entry wall is made from decommissioned street signs. Photo by author, 2015.

This station receives about
1 million pounds of material every day.

Figure 4.3. The transfer station includes an overlook onto the tipping floor; the glass wall has text bubbles with facts about the station and waste system. Photo by author, 2015.

reveal that 2.5 million water bottles are tossed into the garbage every hour in the United States, and that just two water bottles contain enough raw material to generate all of the polyester fiber needed for a baseball cap.

The renovation of the second location, the North Transfer Station in the Wallingford neighborhood, was completed in 2016. Incorporating lessons from the redesign and upgrade of the first station, it is equally representative of the city's ambitions, if not more so. It can accommodate dozens of different material types and includes community space that has nothing to do with waste management, including a park, a basketball court, and a playground. These civic uses help to normalize contact with material management at the urban scale.[56]

Seattle's defiant wasteway endured and reinforced itself over decades, supported by a political appetite for experimentation through six mayors over ten administrations. The wasteway has yielded a number of successful programs and strategies, transformed the roles of citizens and the state, and fundamentally redefined the project of waste management in Seattle. The wasteway nonetheless still operates within the WRWR. When waste managers have attempted something that is too radical, the system has been disciplined by the actors and institutions that maintain the regime.

One example of regime discipline occurred when Seattle attempted to reduce waste generation by regulating the distribution of telephone books. The city created an opt-out program through which residents could choose not to receive a telephone book. Telephone book publishers that ignored opt-out requests were fined by the city to cover the costs of recycling the heavy and unwanted material. Within the first year of the program, 20 percent of Seattle's households and businesses signed up for the program, which resulted in a savings of 375 tons from the recycling stream.[57] Telephone book publishers reacted quickly. In 2014, they sued the city, arguing that the program was infringing on their right to free speech. The initial court proceedings found in favor of the city. But upon appeal, the judge agreed with the publishers, finding no difference between commercial and individual speech. The city had to pay $400,000 for court costs, and the opt-out program became optional for the companies. As Tim Croll described the new program, "We'll take your name down and pass it on, but frankly whether [the telephone book distributors] pay attention to you or not is up to them."[58]

Seattle's waste managers felt that the opt-out program could be an easy win, because it targeted a product that no one really wanted in the first place. But even so, regulating the waste production side of the waste regime proved to be too radical. The legal arguments underpinning the decision remain controversial, but they reinforced the power of producers to control material flows, and the decision forced Seattle to develop a weaker opt-out program with no enforcement.[59]

The failure of the opt-out program demonstrated what happens when Seattle's wasteway pushes too aggressively against the waste regime. It also illuminated two additional aspects of the defiant wasteway: the politics of who initiates new programs, and the adaptive capacity of the Solid Waste Utility. In terms of program origination, some initiatives are devised by Utility staff based on planning documents, public interest, or other motivations. But programs are also sometimes motivated by politicians. The telephone book program was devised and promoted by City Councillor Mike O'Brien, a former Sierra Club activist. From the Utility's point of view, telephone books were not a priority item. They produced "maybe a thousand tons [of waste] a year," Croll said. "[And] of course history and technology are driving these things out [anyway]."[60] But it was an important issue for Councillor O'Brien, and so the Utility supported it. The Utility leadership viewed the relationship with O'Brien as a productive partnership: "We scratched his back and he scratched ours. He supported our stuff," Croll said.[61]

This ability to seize "policy windows," but also to make sure they are aligned with a larger goal, contributed to the momentum toward waste diversion and reduction.[62] The Utility contextualized moves like the telephone book opt-out within a larger planning framework that included higher-priority, higher-tonnage items. In the case of the telephone books, the Utility was then subsequently able to get O'Brien's support to ban Styrofoam take-away containers, a controversial provision that had been on the table for decades, and which the city council finally passed in 2008.[63] Styrofoam is a more challenging and problematic material than telephone book paper: it is not recyclable; its use and production release toxins, ozone-depleting chemicals, and warming compounds; and it is very slow to degrade, meaning it lasts in the environment for a very long time. Ultimately, banning single-use Styrofoam containers is more environmentally significant than reducing the distribution of telephone books. The presence of strong institutions, and, of course, politically savvy technocrats, allowed the failed first initiative to eventually support the second.

A second key failure occurred in 2010 as it was becoming increasingly difficult to shave down the tonnage going to the landfill. As an experiment, the Utility piloted a program of biweekly garbage pickup. Utility leaders believed that if garbage was collected less frequently than organics, it would further incentivize food scrap composting.[64] Throughout the pilot program the Utility kept in close touch with the eight hundred families that were included: they communicated with families prior to beginning the program, and they then used a variety of techniques to gather feedback about the families' experiences with the program. More than a third of the participants ultimately recommended against expanding the program citywide, and a newly elected mayor shelved the program.[65] This willingness to experiment when the outcome is unknown, and the institutional capacity for feedback, communication, and monitoring, were all key factors in the transformation described in chapter 3.

These same factors also constitute key elements of the defiant wasteway. Even while discussing the failure of the biweekly garbage collection program, Tim Croll was already planning for other experiments, such as curbside recycling of construction debris.[66] And, while Mayor Michael McGinn may not have been ready to take on the political challenge so early in his tenure, the *Seattle Times*, which had not always been on the side of experimentation, published an editorial arguing that it was worth figuring out the details of biweekly collection. If the Utility could resolve increased recycling contamination and anxiety about rodents, and allow for customers with special circumstances, it would help the city meet its goal of 70 percent waste diversion.[67] The editorial indicated a degree of public buy-in with the larger system goals, a nontrivial factor as the path forward got steeper. Also, the city has proven again and again that things that seem impossible one year—such as mandatory recycling in 1988—can become politically viable with the passage of time. As changes become normalized, the social setting of garbage management changes; people's expectations change. What is controversial at first eventually becomes common sense.

A third key example of WRWR discipline followed the city's adoption of mandatory food scrap composting. As noted previously, Seattle instituted mandatory food scrap composting in 2015 after a decade-long period of voluntary source separation. In 2014, Utility leaders convinced the new mayor, Ed Murray, and the city council that mandatory food scrap composting was the best option for meeting the 2015 60 percent diversion rate goal.[68] The city council passed an ordinance requiring residents to discard food scraps in their yard waste bins. The Utility planned a series of warnings and fines for residents to be implemented after a grace period, just as when recycling became mandatory years earlier. There was very little protest when mandatory composting was enacted; for the most part, residents in single-family homes had already been separating food scraps. But there was still a lot of food going to the landfill, and the Utility wanted to improve composting participation across the city.

The Utility attempted to boost participation by randomly examining the garbage in people's bins. Residents who continued to dump food scraps in the garbage were subject to warnings and then small fines. The warnings consisted of light pink flyers—a noticeable but not alarming color—affixed to bins; the flyers read, "It's not garbage anymore," and included a reminder of the new rules.

The messaging supported the process of redefinition that had started decades earlier with yard waste and recyclables. The message was that food scraps were not trash, but a valuable resource: they could help restore nutrients to the soil rather than being left to release methane in a landfill. While most Seattle residents were unconcerned about the policy change, a very small minority actively opposed it when the city began enforcement. A group of eight residents enlisted the help of the Pacific Legal Foundation (PLF), a nonprofit organization with a self-stated mission to protect "private property rights, individual liberty, free enterprise, limited government, and a balanced approach to environmental protection."[69] The PLF sued the city, arguing that it was unconstitutional in Washington State to snoop through residents' garbage.

The core issue was privacy. As one resident told a local paper, "The city has no right to know what I eat and drink. You can get all this information about people from their garbage."[70] This framing of garbage—that it is personal, private, and taboo—reflects Victorian and Sanitary City ideologies of waste. But it is also an ideological framing designed to advance contemporary neoliberal political agendas that have nothing to do with trash or the environment. The idea that waste is personal and private, and thus unexaminable, dovetails neatly with the mission of the WRWR, which depends on waste being unseen and unexamined.

The city defended its policy, arguing that waste haulers were doing cursory and random visual inspections, not unlike what they did for recycling; they were not systematically pawing through garbage gathering secrets and meting out fines. The judge ultimately determined that the city had failed to describe how it could possibly identify compostable food scraps in the trash without thorough inspection. The city was, in the words of the attorney from the PLF, "overstepping the line."[71] It was not only violating a limit on state power enshrined in Washington State's constitution, but symbolically transgressing between the realms of waste and not waste, and thus threatening the stability of the waste regime.

In this case, an ideologically driven nonprofit organization, residents upholding deeply held social values, and a supportive judiciary disciplined Seattle's wasteway. News outlets around the country, including the *New*

York Times and Fox News, picked up the story.[72] The widespread interest in the case highlights the powerful values and ideologies buried in our trash cans and shows how waste regime discipline can be broadcast to other cities, constraining the possibilities for municipal waste management nationwide.

Even though the decision forced Seattle's waste managers to reevaluate enforcement strategies, the Utility responded with the characteristic aplomb of the defiant wasteway. It released a statement expressing pleasure that "the court's ruling recognizes the City's ability to regulate what goes into trash cans to address conservation and safety needs. 'Plain view' monitoring for dangerous items is vital to protecting worker and public safety.... We will study the ruling and determine what changes we need to make in the program and the City ordinance."[73] In the aftermath of the decision, the Utility continued to invest heavily in education and communication, and in even one-on-one coaching in the commercial sector. This was just the latest in a long series of successes and setbacks, and the Utility showed no signs of slowing.

Discipline to the waste system comes from industry, from political and legal apparatuses that support industry, and from residents whose expectations are still shaped by deeply held constructs of order, privacy, convenience, and cleanliness (ideologies often actively reinforced and promoted by industry). As anthropologist Mary Douglas has shown, these constructs are fundamental to people's understandings of themselves in the world, and the maintenance of order and cleanliness in particular are sacred values. She also pointed out, however, that while they may appear to be timeless and absolute, "there is every reason to believe [that these ideas] are sensitive to change. The same impulse to impose order which brings them into existence can be supposed to be continually modifying or enriching them."[74] And this has certainly been the case in Seattle. Mandatory recycling was once viewed as invasive and impolitic, but residents and politicians eventually accepted it. Separating organic waste is generally viewed as a nuisance—and somewhat icky—but many Seattle residents do it habitually, and even feel unease in places that do not offer a curbside composting service, even while a vocal few have pushed back against enforcement of the law.

Seattle's defiant wasteway is constantly being cultivated by elected and nonelected officials who are willing to experiment, test, reset, and try again. Any one of these failures could have crippled a system that was not so institutionally reinforced. In Seattle, ambitious goals, a framework for regular monitoring, and a shared belief in waste reduction among managers, policymakers, and stakeholders have reinforced momentum toward reduced reliance on garbage disposal.

The Defiant Wasteway as Resistance

After their crises in the 1980s, Boston's and Seattle's waste systems evolved within the framework of their wasteways. While Boston hewed closely to the weak recycling waste regime by maintaining a system that upheld Sanitary City logic and producer-disposer industry interests, Seattle carved a path of transformation and resistance that has endured, even as it has been disciplined.

The key shifts in Seattle were institutional: formalization of goals; broadening expertise among waste managers; and the Utility's capacity for inclusive planning, outreach, and community feedback and education. These changes bucked engineering trends as well as the trends of neoliberal governance that drove the deskilling and hollowing out of the state. Once public-sector waste managers were no longer simply disposal experts, the Utility developed other sets of skills that focused as much on research and communication as on the science and technology of materials management. Seattle's institutional changes produced the core elements of the defiant wasteway: experimentation, learning from failures, and patience to see through long-arc social and infrastructural change. The defiant wasteway allows for change that would not otherwise be compatible with local electoral politics or the neoliberalization of local governance. It has ensured that the public and the city's leadership have continued to buy in to the city's waste system even when they have objected to specific programs. Seattle's wasteway created its own momentum, its own internal logic, and the resulting system is resilient enough to continue to resist broader national and global regimes, even when disciplined. In short, the defiant wasteway functions as a form of resistance challenging the powerful interests that maintain the WRWR.

CHAPTER 5

Resisting Garbage

Seattle and Boston, along with other cities across the country, experienced declines in regional landfill capacity in the late 1970s and early 1980s. Waste managers and elected officials in both cities gravitated toward space-efficient waste-to-energy incinerators to solve what they understood as a waste disposal problem. But, in both cities, the incinerator proposals failed. In the planning processes that unfolded, each city established a wasteway that shaped decisions and actions for decades to come. In Seattle, an open public discourse, driven by inclusive planning, an engaged citizenry, regional infrastructure politics, and escalating costs, changed civic leaders' understanding of the garbage problem. Stakeholders redefined waste temporally, materially, and spatially. They gave waste new form and new meanings, transforming trash into a stream of potential resources that required attention and stewardship. In the process they transformed the roles of citizens and the state that had dominated since the Sanitary City era. Seattle residents and businesses became active waste managers with increased responsibility, not only for paying for what they wasted, but also for managing increasingly demanding processes of waste separation. They gained expertise that waste planners valued, and decision-makers took account of their views and input. Citizens, in essence, became partners in the project of material stewardship and helped the city government reach their collective goals. The government, still responsible for supporting residents with convenient and functional infrastructure, was essentially recast in the role of a citizen: it became one stakeholder among many, with all parties taking on the same responsibilities for careful consumption. City officials, elected and appointed, became educators, listeners, and stewards.

Over the ensuing decades, these changes cohered into a wasteway that defied the weak recycling waste regime, chipping away at the disposal paradigm that had dominated the extraction-manufacturing-consumption-waste chain since the nineteenth century. The defiant wasteway worked through programmatic innovation that stemmed from deeper institutional changes. Within Seattle's defiant wasteway, waste was not defined as a disposal problem, as is typical in the WRWR, but instead as a stream of resources with unique properties and possibilities that should be stewarded. Technical decisions about how to manage waste were made through processes that included many voices and forms of expertise, which allowed both for widespread buy-in and for solutions that would not have come from sanitation professionals alone.

Change in Seattle was incremental, and the system still relies on a limited set of tools available within the structure of the WRWR — that is, individual behaviors and municipal powers. But the approach promotes resistance to WRWR demands, and the resulting system looks and functions in a very different way from waste systems, like Boston's, that faithfully implement the weak recycling waste regime.

Boston also rejected an incinerator proposal, but the decision largely resulted from a power struggle between city and state elected officials and the narrow set of interests and varieties of expertise that controlled the process. The product was a compliant wasteway that operated neatly within the bounds of the WRWR. Waste remained an abstraction, a thing to be disposed of, and waste managers continued to prioritize hygienic and efficient collection. With a lack of resources for planning and experimentation, a very limited public process, and a continued emphasis on traditional waste management expertise, the resulting system failed even to meet the national average recycling rate.

The Practical Extent of Seattle's Transformation

Despite Seattle's many achievements, the fact remains that the city sends a mile-long, double-stacked freight train full of garbage to a landfill in Eastern Oregon every day. Waste managers in the city are well aware that their programs have yet to truly tackle the dominant paradigm of consumption and disposability. Although the city has taken measurable steps toward waste reduction, and the city's recycling rate is impressive — all the more so because it includes the commercial sector, which most cities do not even measure — the system necessarily remains a (somewhat uncooperative) creature of the WRWR.

Furthermore, the city's waste diversion outcomes remain uneven. In 2018, the city had a total recycling rate of 56.5 percent; residents of single-family homes, who generate 30 percent of the city's waste stream, achieved a 72.6 percent recycling rate. But the city has not managed to achieve the same level of participation in multifamily buildings. The multifamily sector, which generates 10 percent of the city's waste stream, had only managed to reach a 36.4 percent recycling rate by 2018.[1] Because multifamily housing has been growing rapidly, this is likely where future waste-stream growth will originate.

The divide between single-family and multifamily recycling rates is largely the product of technical constraints. Apartment dwellers do not sign up individually for garbage service, and therefore do not have the same cost incentives as single-family home dwellers. Enforcement is more difficult when waste streams from multiple households are comingled. Residents in apartments often have less space for separating and storing multiple material types. And apartment complexes are rarely designed to accommodate the separate storage and collection of three distinct discard streams.[2]

The divide between the single-family and commercial sectors is less stark, but more consequential. In 2018, the commercial sector reported a slight decrease in waste generation and a marginal increase in recycling and composting, with a 65.1 percent recycling rate. Given that commercial waste represents 48 percent of the city's municipal waste stream, improving performance is a priority for the Utility.[3]

Overall that same year, diversion rates dipped by 0.5 percent in Seattle. Among possible reasons, the Utility suggested that many recyclable materials are becoming lighter than in the past, which reduces the tonnage even as volumes continue to increase. Additionally, the city has invested a considerable amount of effort and money in recycling education. The Utility speculates that wishful recycling—the phenomenon of recycling anything you hope might be recyclable—is decreasing, in part as a result of the city's "When in doubt throw it out" campaign and other educational efforts. This means an improvement in the material quality of recyclables despite a reduction in overall tonnage. Finally, the Utility speculates that continuing press about China's decision to increase the quality requirement for recyclables from the United States may have dampened public faith in recycling. This phenomenon may have been further exacerbated by the coronavirus pandemic, which led to national press about stress on municipal recycling across the United States.[4]

Seattle recycles and composts more material than almost any other American city—but within the WRWR, recycling is not necessarily a signal of success. Critics note, for example, that recycling actually encourages consumption by easing the consciences of consumers, and that it addresses waste only at the very end of the production chain, ignoring the whole system of waste-making from resource extraction through manufacturing and transport. Moreover, processing recycled materials requires huge amounts of energy and degrades material quality. Recycling, argue critics, is a ploy used by industry to displace the responsibility for environmental action onto individuals and municipal governments and to deflect regulatory attention from the corporate machinery of consumption and waste-making. Finally, municipal programs emphasize materials for which recycling provides minimal environmental benefit, ignoring other materials that could be more efficiently recycled.[5] And these are just a few of the many arguments against the efficacy of recycling as a solution to the problems of garbage. Seattle's success may be notable, but its recycling rate hardly represents a solution to the many socio-environmental crises we face.

The limits of Seattle's progress are even more stark when put into the context of the rest of the country: only a handful of cities even come close to achieving similar rates of reduction and diversion, and even fewer have launched the whole-scale redefinition of solid waste management that Seattle has undertaken. Representatives of cities from across the country, and indeed from around the world, frequently visit Seattle and send inquiries about Seattle's programs. In response, the city's 2011 solid waste plan update was structured explicitly to serve as a "resource document" for others as much as a guiding plan for the city.[6] Visitors, however, most frequently request information exclusively about the Seattle waste and recycling programs; according to Tim Croll, the solid waste head through 2016, when waste managers hear about Seattle's programs the response is often, "Wow, you're really doing that. Is that true? Oh, I don't think we could ever do that."[7] As Croll suggests, this programmatic approach misses the point about Seattle: it is not any one individual program that yields the city's nearly 60 percent diversion rate. It is the multiple institutions of the defiant wasteway as they have matured over time. Cities that simply try to replicate Seattle's recycling program without the underlying institutional transformations will only reinforce the WRWR.

In short, it is easy to critique Seattle's success because the WRWR is still intact. So how can we understand Seattle's evolution, if not as an uncomplicated success story?

The key is its defiant wasteway, not just its impressive recycling rate. In material terms, the defiant wasteway in Seattle has actually succeeded in reducing waste generation—a remarkable feat, given the lack of support from larger-scale policy. Between 2008 and 2018, waste generation did not keep pace with population growth. The population grew 5 percent during that decade, and median household income rose nearly 30 percent over the same period. But waste generation only increased by about 1 percent. The per capita waste generation rate actually *decreased* by 21 percent over those ten years, to 2.16 pounds per person per day.[8] People in Seattle throw less away than they used to. And they throw away about half of what the average American discards daily.[9]

Despite Seattle's gains, the Utility is not content with its progress. The system's aggressive goals and support for experimentation mean that the city's waste managers have moved well beyond low-hanging fruit to address more challenging barriers to recycling and reduction, including in the most challenging segments of the market. For example, the Utility recently was granted a formal role in design approvals for new multifamily buildings; its representatives can advise developers during the design process to ensure sufficient space in these projects for separating and storing multiple material streams.[10] Seattle has also not been afraid to experiment with regulating consumption. Its 2011 waste management plan update contains an entire chapter on waste prevention that emphasizes structural changes, such as product stewardship and extended producer responsibility, that depart meaningfully from neoliberal dependence on individual behaviors.[11] The city was one of the first in the nation to ban plastic grocery bags and foam takeout containers. These bans do not regulate production, but do slightly alter the structural landscape of consumption, and they are a step upstream from relying exclusively on individual consumer choice that is not infrastructurally supported. There are many critiques of such efforts to regulate plastic consumption. Nevertheless, these steps are a radical act because they regulate the products available to consumers in the WRWR.

Seattle's real reduction in waste generation, its commitment to continued progress, and its willingness to regulate consumption are all products of the defiant wasteway. They are evidence of a reorganized system that actively resists the WRWR, using the tools available at the scale of local government.

The Wasteways Framework as an Analytical Tool

Defiant wasteways like Seattle's, and compliant wasteways like Boston's, will look different in each municipal setting. Every city system has evolved along its own path, with its own particular politics, its own personalities, its own infrastructural priorities. There are cities that divert even less than

Boston does; there are cities, like Austin, that have been reforming norms and institutions but have yet to achieve the same waste diversion and reduction success as Seattle. There are many, many cities in between, including Los Angeles, that have set ambitious goals for reduction but are still trapped inside of waste governance institutions built by Sanitary City and WRWR priorities.[12]

The wasteways framework reveals distinctions between urban waste systems that are not easily visible in more standard metrics of comparison. For instance, as of 2019, Houston and New York had similar residential recycling rates, 19 percent and 21 percent, respectively—neither one so different from Boston's rate.[13] A wasteways analysis, though, reveals significant differences between the systems. New York makes garbage statistics easily available online, creating transparency about its systems and opening up its data to all who are interested. The city also invested in an ambitious residential curbside composting pilot project and launched a marketing campaign to let residents know that, despite changes in global recycling markets, the city's recycling efforts are still effective.[14]

A recent effort to improve waste management in Houston reveals very different priorities. Houston considered moving to a form of single-bin collection in which all organics, recyclables, and waste materials would be comingled and then processed through high-tech, back-end technologies such as mechanical sorting and gasification. The project, called "One Bin for All" and funded by Bloomberg Philanthropies, promised to use technology to "leapfrog" to higher recycling rates.[15] In a report on the One Bin project, city officials noted that they had learned that public participation was necessary to bridge "the lack of a common lexicon between waste experts and the public, which will result in confusion and a loss of time."[16] In other words, they understood that a stakeholder process was necessary in order to successfully convince the public that expert decisions were correct. The One Bin process and the city's approach to stakeholder engagement together would have cemented perceptions of the respective roles of city officials and citizens: the former as experts and service providers, the latter as consumers and waste generators with no relevant knowledge or experience. The approach would have alienated people further from their discards: they no longer would have had to think about the material properties of their waste or their disposal practices at all, an approach that would have allowed them to consume anything and everything without a second thought. The single-bin system would have relied on back-end proprietary sorting technologies, which would ultimately serve disposal industry interests. Houston's One Bin proposal is what "progress" looks like in the WRWR.[17]

New York and Houston have similar recycling rates, but the underlying cultures of waste governance—and their resulting wasteways—are quite different. New York is not known for transparent urban governance, and its sanitation system is famously maligned. Nevertheless, its public data, emphasis on organics, and attempts to communicate through public campaigns indicate the beginnings of a defiant wasteway. Houston's One Bin program, in contrast, was poised to double down on WRWR relations and processes. Wasteways analysis helps to distinguish cities that are promoting a new regime from cities that are propping up the old one. It may also provide some clues as to how resilient a city will be if and when the regime changes.

The WRWR under Stress

Understanding municipal wasteways is useful for a few reasons. One is that cities have become a critical scale for environmental governance.[18] Waste management is a crucial link between cities and the global extraction-manufacturing-consumption-waste chain. Looking at recycling rates alone masks a great deal of innovation and radical action that is happening at the city level as municipal governments shift into new modes of environmental governance. Wasteways analysis can reveal the kinds of institutional innovations occurring at the local level that challenge resource regimes, material flows, and the relations of the consumption economy.

In addition to insights about environmental governance, wasteways can also show us how city governments are responding to what looks a lot like the next garbage crisis. Popular concern about garbage is exploding, with particular anxiety about plastics. In 2018, *National Geographic* launched a series called "Planet or Plastic" as part of a multiyear effort to "raise awareness about the global plastic waste crisis."[19] Plastic pollution in oceans, rivers, soil—and even human bodies—received escalating mainstream media attention throughout the 2010s.[20] Popular media has been paying attention to a rise in plastic production, a storyline that organizations like Keep America Beautiful suppressed for years. Documentaries and investigative journalists have highlighted massive investments in plastics by the fossil fuel industry as the latter has attempted to ensure future profits in the face of competition from renewable energy and profitably soak up surplus natural gas production.[21]

Concern about plastic is also manifesting in popular culture. The first few episodes of the show *Veep*, for example, featured a vice-presidential team fretting about which big oil executive to place on a green jobs task force; the vice president's aides needed oil to satisfy senators in the pocket of fossil fuel, but had to find just the right person so as not to alienate the environmentalists on the task force whose primary concern was plastic. In a humorous stunt in the first episode, the veep herself experiments with cornstarch utensils, imagining that if federal buildings start stocking them instead of plastic, it could be a big win for her. Unfortunately, the cornstarch spoons couldn't stand up to a cup of coffee—let alone the satirical politics of the fictional Capitol Hill.

Just as the rise in disposable packaging created the litter crisis in the 1950s, and landfill closures spurred the garbage crisis of the 1980s, plastics are now at the center of a new waste crisis. *Veep* showcased popular anxiety about the structural power of plastic and the oil industry through satire, and it placed the debate within the wonky world of DC insider-politics. But plastic anxiety is no longer contained in wonky science or policy circles. Documentary films and long-form narrative exposés are also abundant. PBS aired an hour-long special called *The Plastic Problem* in 2019. Indeed, dozens of films about garbage and plastic were produced between 2010 and 2020. To name just a few: *A Plastic Ocean* (2016), *Wasteland* (2010), *Plastic China* (2016), *Garbage Island* (2008), *Toxic Garbage Island* (2014), *Plasticized* (2011), *Plastic Paradise* (2013), *Plastic Planet* (2009), *Bag It* (2011), *Albatross* (2018), *Addicted to Plastic* (2008), and the 2020 follow-up to the 2007 film *Story of Stuff* called *The Problem with Plastic.*

The media explosion around plastic can be interpreted in many ways, but there is one clear takeaway: people are worried about plastic. This concern is not unreasonable. The EPA refers to plastics as "a rapidly growing segment of municipal solid waste," which may be somewhat of an understatement. In 1960, the United States generated some 390,000 tons of plastic waste. That number had grown 100 times by 2017.[22] The United Nations estimated in 2019 that the 300 million tons of plastic produced annually around the world weighs nearly as much as the total human population of the earth.[23] In the United States, only about 8 percent of plastic is recycled, according to EPA estimates;[24] globally, about 9 percent is recycled annually.[25] According to estimates, about 12 percent of the plastic ever produced has been incinerated. And because plastic does not biodegrade, low recycling and incineration rates mean that nearly *all plastic ever made* still exists in some form on earth, much of it in landfills, and increasingly, in oceans. The Ocean Conservancy estimates that plastic bags alone kill 1 million sea birds and 100,000 other animals every year.[26] An increasing amount of plastic can now also be found in the food we eat and in our bodies.[27] Plastic has become so pervasive that scientists expect it will eventually become the geologic indicator of the Anthropocene.[28]

When litter exploded in the American landscape in 1959, Vermont passed a landmark ban on disposable bottles. Industry organized immediately to contravene regulation of the production of disposable packaging and single-use goods, giving rise to organizations like Keep America Beautiful and the Recycling Partnership. Working in the interests of manufacturing and waste management industries, these lobbying organizations successfully changed the public perception of waste and redirected regulatory energy toward local recycling programs. The coalition's response to the perception of a litter crisis in the 1950s and the growing environmental movement transformed the Sanitary City into the weak recycling waste regime. This same producer-disposer coalition continues to reinforce the WRWR by ensuring that basic definitions of waste, the parties responsible for waste, and the options for managing waste remain limited and stable. But the WRWR systems and infrastructures of disposal are not adequately managing the ongoing deluge of plastic, and this this failure could destabilize the WRWR.

Predictably, waste and manufacturing industries are organizing to prevent plastics regulation. Plastics lobbyists have become adept at encouraging state legislatures to pass preemption laws that make it illegal for municipal governments to regulate plastic consumption. As of 2020, such laws had been passed in fourteen states: Arizona, Colorado, Florida, Idaho, Iowa, Michigan, Minnesota, Mississippi, Missouri, North Dakota, Oklahoma, Tennessee, Texas, Wisconsin. To compel state legislatures to act in the interests of industry, lobbying groups make a host of arguments to defend infinite and freely available single-use plastic items. In doing so, they mobilize social justice rhetoric (e.g., a bag fee amounts to a regressive tax) and environmental arguments (e.g., paper bags are heavier and require more fuel for shipping).[29] In the wake of the coronavirus pandemic, industry launched concerted lobbying campaigns against reusable consumer goods, arguing, without evidence, that they spread the virus and that plastic saves lives.[30] These techniques of opportunism and co-optation are straight out of the WRWR playbook; they mirror almost exactly the tactics of Keep America Beautiful in the 1960s.

Throughout garbage discourse over the past decade, the coalition of producers and manufacturers that has had the greatest stake in maintaining the WRWR has held fast to core positions: plastic is only a problem if it is not recycled properly, and responsibility for managing plastic lies with consumers. The American Progressive Bag Alliance, which runs the "Bag the Ban" website, argues that "recycling is the real solution to waste and litter in our environment." Plastic bags, argues the alliance, are "100 percent recyclable." Just "bring your plastic bags, sacks, and wraps to participating retail stores and drop them into plastic bag recycling bins," they suggest.[31] This framing puts responsibility entirely on consumers (and "participating retail stores") and sidesteps the problematic fact that, when recycled, plastic polymers degrade. They cannot be infinitely cycled into new plastic products. Recycled plastic bags still require virgin materials to maintain strength and quality. This logic extends to the entire plastics industry.[32]

In discussing an enormous Royal Dutch Shell plastic pellet manufacturing plant outside of Pittsburgh, spokeswoman Hilary Mercier told the *New York Times* that plastic is only problematic when it is disposed of improperly. Royal Dutch Shell, she emphasized, "passionately believe[s] in recycling."[33] This "passionate" belief, as we have seen, is quite limited. Industrial proponents of the WRWR only believe in recycling to the extent that its costs fall elsewhere. Bottle bills, for example, are extremely effective at promoting recycling. But they also force producers and retailers to organize and pay for recycling. Not surprisingly, the beverage and retail industries organize fiercely to oppose any proposed new or expanded bottle bill.[34] In short, a coalition of plastic-reliant industries, including the fossil fuel industry itself, marshals predictable rhetoric to deflect regulation that would disrupt infinite production, consumption, and disposal of plastic—that would disrupt, in other words, the WRWR.

The disposal side of the regime is also doing its part to control garbage discourse and shore up the WRWR in the face of public anxiety and critique. In the summer of 2019, a friend texted me a photo of a Waste Management, Inc. (WM), garbage truck trundling through her neighborhood in Chicago. The green truck in the photo was sparkling clean, and the driver regarded the camera suspiciously from the open truck window. Blazoned on the side of the truck was this provocation: "What if nothing was considered waste? It begins right here." WM knows that people are concerned about garbage—and plastic, in particular. From the company's perspective, this concern is dangerous. What if people mobilized and successfully brought about a massive reduction in garbage? What if consumers stopped using disposable items? What if the entire economy changed to such a degree that material throughput shrank? WM would lose business. The question on the truck—"What if nothing was considered waste?"—is a ploy to control the narrative by co-opting the problem-framing process. The motto suggests that the problem lies in *the naming*, not in wasteful behaviors, overproduction, toxic materials, or extractive economic structures. "It begins right here" tells us that whatever the problem is, it's not ours. The corporation, acting judiciously somewhere beyond the end of the pipe, will take care of it. The renaming of garbage keeps the emphasis at the end of the chain; it is, really, an *unprobleming* of garbage. WM's role in the process is more than just rhetorical, and it extends far beyond signs on garbage trucks. Waste Management, alongside Coca-Cola and Amazon and dozens of other corporate producers, disposers, and distributers, funds Keep America Beautiful and the Recycling Partnership.[35] The message on the truck contributes to public goodwill and quietly shapes public perceptions of garbage. Behind the scenes, WM lobbies directly and through nonprofit partners for policies that protect the regime, and thus WM's interests.

Unlike the 1950s, however, when industry pushback decisively turned the tide against upstream regulation, until the 2019 coronavirus pandemic, plastics laws were only growing in scope and scale, affecting larger and larger portions of the population across key US markets. Despite regulatory silence from the federal government, hundreds of city and state governments were considering, or had already passed, laws banning or limiting the distribution of single-use items such as plastic straws and shopping bags, foam take-out containers, and more. According to the Surfrider Foundation, a grassroots environmental organization, for example, there were 471 local plastic bag ordinances in twenty-eight states as of the fall of 2019. Fifty-eight of those ordinances were passed just during the summer and fall of 2019. The impact of these ordinances was uneven and depended a great deal on how they were implemented. But the best of these programs had documented an impact at the local level. Both Washington, DC, and San Jose, California, for example, reported nearly 90 percent reductions in plastic bag pollution in storm drains, and many studies have shown that imposing even tiny fees on all bags significantly decreases the number of bags used.[36]

In addition to regulation of single-use plastics, a simultaneous explosion is taking place in right-to-repair laws and extended producer responsibility (EPR) legislation.[37] More than twenty states were considering some form of a right-to-repair law by 2019, and dozens of states and city councils had implemented EPR programs for everything from house paint to mobile phones. Right-to-repair and EPR laws, which demand the production of repairable consumer goods, may be the most effective available tool against planned obsolescence and disposability, as they force corporations to take responsibility for the waste they manufacture and to prevent waste from being generated in the first place.

The question remains, however, whether the hundreds of garbage-resisting, plastic-reducing, right-to-repair, and EPR ordinances now on the books in American cities and towns are part of defiant wasteways that are institutionally reorganized to resist the WRWR, or simply easy wins targeting marginal material streams that appeal to local governments that have no choice but to at least look "busy" in the fight against the climate emergency and other environmental catastrophes. There are certainly many critics charging that straws and bags are distractions, like recycling itself, because they keep popular and regulatory attention from recognizing more pressing problems.[38]

But if plastic is oil (which it is), then regulating its consumption and forcing producer responsibility would seem to be a signal that stakeholders representing a nontraditional set of interests are engaging in policy-making at the local level. This is especially true in places like Laredo, Texas, where an unusual coalition of cattle ranchers and environmentalists (not unlike the coalition that initially supported Vermont's 1959 disposable bottle ban) supported a plastic bag ban. Texas eventually succeeded in preempting Laredo's plastic bag ordinance, but not before an expensive lawsuit involving the city.[39] When cattle ranchers intervene on the side of less plastic, it is an inversion of typical political alliances that suggests the crisis is deep. When cattle ranchers participate in solid waste policy-making, it suggests a defiant wasteway.

Resisting garbage is a radical act even when done at the scale of the individual. It means reconsidering one's wants and needs, developing new habits, and opting out of many normal, naturalized modes of consumption. But individual action is not an option for many people, whether because of lack of money or time, because of disabilities, or because of living in a setting that simply does not offer alternatives. Individual action is clearly not enough, especially for the many for whom it is impossible. When scaled up to the municipal or even the federal government, resisting garbage—even a seemingly small item, like a plastic bag—means reconstructing fundamental assumptions about how the economy should function and for whom. Banning plastic bags will not avert the climate crisis. But resisting garbage is a shift in priorities, a challenge to the whole damaging, unequal system of production and consumption. We have so few examples of municipal policy that even attempts to intervene in the global production-consumption-waste chain at all; in fact, usually municipal policy is designed to encourage more development and more consumption. In Seattle, redefining garbage was the first step on a long path of transformation and eventually resistance, and it could be a first step in other places as well.

Seattle and Boston offer two views of resistance and acquiescence within the constraints of the WRWR and demonstrate how those orientations can translate into decades of policy and service outcomes. Seattle's defiant wasteway resists garbage in many ways, starting with the fact that its waste management system defines very little as actual trash. It resists garbage materially by recycling and composting. It resists garbage structurally by limiting what products are available to consume and by promoting, and in some cases requiring, reuse, repair, and extended producer responsibility. It resists garbage by partnering with citizens and by creating a public coalition of garbage resistance rather than leaving individuals on their own with an impossible challenge.

The world has changed since 1980, and the nested crises of climate and plastic are altering the calculus of municipal policy. Even Boston is reconsidering its orientation to the WRWR. In 2018, the city launched an inclusive and thoughtful Zero Waste planning process; although the scope was limited, this decision nevertheless represented a massive shift from prior processes. Waste management, recycling, Zero Waste, environmental justice, labor experts, and interested residents all had multiple opportunities to view and weigh in on the plan.[40] The resulting plan included an ambitious goal to increase diversion to 80 percent by 2035. To get there, it recommended not only pursuing improvements in composting and conventional recycling, but also bolstering existing channels of reuse and nurturing industries that rely on recycled materials. If Boston succeeds in making these changes, it, too, will be resisting garbage. If other cities do the same, then the assemblage of defiant wasteways will eventually force a regime change, replacing the WRWR with something new. The WRWR cannot be sustained without the willing participation of citizens and city governments.

Seattle offers a small glimpse of what a post-WRWR world could look like. It is not a panacea. But it is a glimmer of hope, a possibility that resisting garbage can make a new world, even as it doesn't solve all of the world's problems. Beyond Seattle, and perhaps, in part, because of Seattle, we are seeing incremental change alongside steady—but not entirely successful—retrenchment. Given the escalating crisis of plastic in the environment and flagging faith in WRWR answers, it is starting to look like resisting garbage is our future.

Acknowledgments

Garbage is a strange thing to obsess over. It is precisely the thing that we are not *supposed* to think about, which is perhaps why I landed on it. I had no idea, when I first began this project (many, many years ago) that it would open so many doors. Garbage, it turns out, is not at all alienating.

My first foray into the study of waste occurred when I was an undergraduate student in David Smiley's urban studies seminar at Barnard College. Since that time, when I was the grateful recipient of David's good-natured encouragement, many people have contributed to my intellectual engagement with trash generally, and this book specifically, in ways both great and small, and all were essential to the development of my ideas.

This project began in earnest when I was at the Massachusetts Institute of Technology working on my PhD. Funding from MIT's Department of Urban Studies and Planning supported multiple trips to Seattle. Eran Ben-Joseph, Larry Vale, Judy Layzer, Jennifer Light, and Gabriella Carolini all nurtured this project from incipient idea to research plan to actual argument. Thank you Eran, especially, for helping to translate that messy process into a job and a book and for remaining a steadfast mentor through the years. I'm not quite sure how to adequately acknowledge Judy Layzer. She is not here to see the final version, but it was her methodological rigor that kept me digging in the archives, making cold calls, and coding endless documents, even as my resolve began to flag, until the full story related in this book became clear. And it was Judy's fundamental optimism about people and planning that helped me identify a narrative of hope and resistance in trash.

Along the way, from the first days of PhD research to a book prospectus, a host of scholars and friends have supported the development of this project. I owe special thanks to my CEOs Lyndsey Rolheiser and Anthony

Vanky. Kian Goh, Aditi Mehta, Elizabeth Walsh, Jordan Howell, and Abby Spinak all provided supportive, challenging, eye-opening insights, sometimes about this research, but more often, and perhaps more importantly, about how the world works. Devanne Brookins once said that my work made garbage seem interesting, and years later, I still regard that as the highest compliment. I am incredibly grateful to Jonathan Krones for endless conversations and questions about trash and for being such a fun and energetic collaborator.

A few dedicated friends and garbologists have been especially generous with reading various drafts and helping me to refine my ideas. Amy Perlmutter shared work, ideas, and a whole Boston network that deepened my understanding of cities, waste, and processes of change. She also read an early version of this book and convinced me to keep working on it. Mieka Ritsema and Barbara Knecht were my first true garbage companions and provided trenchant feedback on key chapters. James Burns read, and then read again, and I'm so grateful for his smart and honest feedback.

Raechel Lutz and participants in the New York Metropolitan Seminar on Environmental History helpfully pushed me toward more intellectual clarity on the Seattle case. Martin Melosi, whose work first revealed to me that garbage was a medium for understanding history, cities, and the world, generously read and critiqued the framing argument and Sanitary City history when it was still a half-baked dissertation chapter.

Many people who actually did the work of shaping waste management policy in Seattle and Boston in the 1980s shared their memories and insights with me in often lengthy interviews. I am deeply grateful for their contributions, and I sincerely hope that this book honors their work and legacies, even—or especially—when I engaged critically with their stories. I am also grateful to the many librarians and archivists, especially Anne Frantilla in Seattle and Marta Crilly in Boston, who spent many, many hours finding documents and refiling them, making sure I found every extant scrap of relevant data. More instrumental, but equally essential, Kimbra and Tyson Wellock shared their home, friendly dogs, home-cooked dinners, library card, friendship, planning connections, and the introductory anecdote in this book.

Most of the work of writing this book took place at Hunter College. I am lucky beyond measure to have earned a space among the faculty in the Department of Urban Policy and Planning. I am especially grateful for Victoria Johnson's mentorship and Yafit Shafer-Sull's trusty research assistance. And a huge thank you to Robert Devens, Lynne Ferguson, and the staff at the University of Texas Press, who sustained momentum on this project through the upheaval of a global pandemic. Special thanks to Kathy Streckfus for her careful attention to all the little details that make up the whole.

As much as I have enjoyed the adventure of researching and writing this book, the work was not only a scholarly exercise and it was not mine alone. Thacher Tiffany's tolerance for my perpetual blustering about trash has helped immensely. His belief that I could do it has been everything. It is for my children, Theo and Ada, and my nieces and nephews, Alden, Wyatt, Ariel, Micah, and Leor, that I dwell on my hopes for our future world and what the path there might look like. Their amusement about the fact that I am writing a book about garbage has also helped me to maintain a sense of humor (sometimes) about the whole thing. My beloved sisters, given and chosen, Margot Pollans and Beatrice Radden Keefe, provided intellectual support, patient listening, and distraction when necessary. Joan and Ed Tiffany have been enthusiastic champions, gracious hosts, and essential babysitters. I would not have been able to write this book without Kongfang Thunyakit, Anika Smith, Emma Crockford, Rosa Genetti, and Sam Thacher.

My parents, Barbara Baum Pollans and Larry Pollans, taught me by example how to trust in the process and stick with projects through the long haul, even when it is not at all clear where those projects are going. For years they have also read and edited bits and pieces of this work, indulged me in endless discussion, brainstormed just the right terminology, listened to me rant, and dutifully resisted garbage in their own lives. This book is lovingly dedicated to them.

Notes

Introduction

1. "2015 Recycling Rate Report," Seattle Public Utilities, July 1, 2016, www.seattle .gov/util/cs/groups/public/@spu/@garbage/documents/webcontent/1_052510.pdf; "Mayor's Performance Report: Boston Public Works Department," Boston Public Works Department, City of Boston, December 2015.

2. Martin Bruckner, Stefan Giljum, Christian Lutz, and Kirsten Svenja Wiebe, "Materials Embodied in International Trade—Global Material Extraction and Consumption Between 1995 and 2005," *Global Environmental Change* 22, no. 3 (August 2012): 568–576, https://doi.org/10.1016/j.gloenvcha.2012.03.011.

3. Some estimates indicate that MSW is less than 10 percent of the total amount of waste from extraction, manufacturing, and consumption processes. Others have suggested that for some sectors, MSW is closer to 40 percent. In any case, MSW is only a fraction of the solid waste generated by economic activity. Max Liboiron, "Solutions to Waste and the Problem of Scalar Mismatches," *Discard Studies* (blog), February 10, 2014, https://discardstudies.com/2014/02/10/solutions-to-waste-and-the-problem -of-scalar-mismatches; Jonathan S. Krones, "Accounting for Non-Hazardous Industrial Waste in the United States" (PhD diss., Massachusetts Institute of Technology, 2016).

4. Oran R. Young, *Resource Regimes: Natural Resources and Social Institutions* (Berkeley: University of California Press, 1982).

5. Zsuzsa Gille, *From the Cult of Waste to the Trash Heap of History: The Politics of Waste in Socialist and Postsocialist Hungary* (Bloomington: Indiana University Press, 2007), 34.

6. Gille, *From the Cult of Waste*, 34.

7. Unless specifically noted otherwise, when I use the term "production" in this text, I am referring to Gille's concept of *waste production* as opposed to a more general concept of economic production.

8. Joel A. Tarr, *The Search for the Ultimate Sink: Urban Pollution in Historical Perspective* (Akron, OH: University of Akron Press, 1996).

9. Gille, *From the Cult of Waste*, 35.

10. Herman E. Daly, "Reconciling the Economics of Social Equity and Environmental Sustainability," *Population and the Environment* 24, no. 1 (September 2002); Nicolas Georgescu-Roegen, *The Entropy Law and the Economic Process* (Cambridge, MA: Harvard University Press, 1971); Giorgos Kallis, *Degrowth* (Newcastle upon Tyne: Agenda, 2018).

11. William Rees, "Carrying Capacity, Globalisation, and the Unsustainable Entanglement of Nations," in *Globalisation, Economic Transition and the Environment*, ed. Philip Lawn (Cheltenham, UK: Edward Elgar, 2013), 50–65, www.elgaronline.com /view/edcoll/9781781951408/9781781951408.00012.xml.

12. "Sustainable Materials Management: The Road Ahead," United States Environmental Protection Agency (EPA hereafter), June 2009, www.epa.gov/smm/sustain able-materials-management-road-ahead.

13. "Global Material Resources Outlook to 2060: Economic Drivers and Environmental Consequences," Organisation for Economic Co-operation and Development (OECD hereafter), October 2018, www.oecd.org/environment/waste/highlights-glo bal-material-resources-outlook-to-2060.pdf.

14. "Material Consumption (Indicator)," OECD, https://doi.org/10.1787/8497 1620-en (accessed May 28, 2020). According to OECD data, about fifteen OECD countries exceed the United States in per capita material consumption. Most are wealthy countries in cold climates, such as Canada, Finland, and Norway.

15. "U.S. Material Use Factsheet," Center for Sustainable Systems, University of Michigan, August 2019, http://css.umich.edu/factsheets/us-material-use-factsheet.

16. US Census Bureau, Census 1970, Census 2000, via American FactFinder, July 2, 2020.

17. "U.S. Material Use Factsheet."

18. Iddo K. Wernick, "Consuming Materials: The American Way," *Technological Forecasting and Social Change* 53 (1996): 111–122.

19. Dietmar Offenhuber, David Lee, Malima I. Wolf, and Santi Phithakkitnukoon, "Putting Matter in Place," *Journal of the American Planning Association* 78, no. 2 (2012): 173–196; Jennifer Gabrys, "The Distancing of Waste: Overconsumption in a Global Economy," in *Confronting Consumption*, ed. Thomas Princen, Michael Maniates, and Ken Conca (Cambridge, MA: MIT Press, 2002), 155–176; Jennifer Gabrys, *Digital Rubbish* (Ann Arbor: University of Michigan Press, 2011); Jennifer Clapp, *Toxic Exports: The Transfer of Hazardous Wastes from Rich to Poor Countries* (Ithaca, NY: Cornell University Press, 2001).

20. Juliet Schor, *Plenitude: The New Economics of True Wealth* (New York: Penguin, 2010); Tim Kasser, "Materialism and Living Well," in *Handbook of Well-Being*, ed. E. Diener, S. Oishi, and L. Tay (Salt Lake City: DEF, 2018); Robert H. Frank, *Luxury Fever: Money and Happiness in an Era of Excess* (Princeton, NJ: Princeton University Press, 2000); Clair Brown, *Buddhist Economics: An Enlightened Approach to the Dismal Science* (New York: Bloomsbury, 2017).

21. Just a few examples: Joseph M. Braun, Sheela Sathyanarayana, and Russ Hauser, "Phthalate Exposure and Children's Health," *Current Opinion in Pediatrics* 25, no. 2 (April 2013): 247–254, https://doi.org/10.1097/MOP.0b013e32835e1eb6; "Phthalates Factsheet," Centers for Disease Control and Prevention, National Biomonitoring Program, April 7, 2017, www.cdc.gov/biomonitoring/Phthalates_FactSheet.html; "Scope

of Risk Evaluation for Cyclic Aliphatic Bromides Cluster," EPA, Office of Chemical Safety and Pollution Prevention, June 2017, www.epa.gov/sites/production/files/2017 -06/documents/hbcd_scope_06-22-17_0.pdf; Susan Freinkel, *Plastic: A Toxic Love Story* (Boston: Houghton Mifflin Harcourt, 2011); Jennifer Gabrys, Gay Hawkins, and Mike Michael, eds., *Accumulation: The Material Politics of Plastic* (Abingdon, Oxon, UK: Routledge, 2013); Alexis Temkin, "Breakfast with a Dose of Roundup?," *EWG* (blog), August 15, 2018, www.ewg.org/childrenshealth/glyphosateincereal; Nicola Davis, "Toxic Chemicals in Household Dust Linked to Cancer and Infertility," *Guardian*, September 14, 2016, www.theguardian.com/science/2016/sep/14/toxic-chemicals -household-dust-health-cancer-infertility; Theo Colborn, Dianne Dumanoski, and John Peterson Myers, *Our Stolen Future: Are We Threatening Our Fertility, Intelligence, and Survival? A Scientific Detective Story* (New York: Dutton, 1996); Janet Nudelman and Connie Engel, "Right to Know: Exposing Toxic Fragrance Chemicals Report," Breast Cancer Prevention Partners, September 2018, www.bcpp.org/resource/right-to -know-exposing-toxic-fragrance-chemicals-report.

22. J. Timmons Roberts and Bradley C. Parks, "Ecologically Unequal Exchange, Ecological Debt, and Climate Justice: The History and Implications of Three Related Ideas for a New Social Movement," *International Journal of Comparative Sociology* 50, no. 3/4 (June 2009): 385–409, https://doi.org/10.1177/0020715209105147; Sharon L. Harlan, David Pellow, J. Timmons Roberts, and Shannon Elizabeth Bell, "Climate Justice and Inequality," in *Climate Justice and Inequality*, ed. Riley E. Dunlap and Robert J. Brulle (New York: Oxford University Press, 2015), 127–163; Julie Sze, *Noxious New York: The Racial Politics of Urban Health and Environmental Justice* (Cambridge, MA: MIT Press, 2007); Robert D. Bullard, *Dumping in Dixie: Race, Class and Environmental Quality* (Boulder: Westview Press, 1994); JoAnn Carmin and Julian Agyeman, *Environmental Inequalities Beyond Borders: Local Perspectives on Global Injustices* (Cambridge, MA: MIT Press, 2011); United Church of Christ, "Toxic Wastes and Race: A National Report on the Racial and Socio-Economic Characteristics of Communities with Hazardous Waste Sites," Commission for Racial Justice, 1987.

23. "Advancing Sustainable Materials Management: 2017 Fact Sheet," EPA, Office of Land and Emergency Management, November 2019, www.epa.gov/sites/produc tion/files/2019-11/documents/2017_facts_and_figures_fact_sheet_final.pdf.

24. US Census Bureau, Census 1960; US Census Bureau, "American Community Survey," 2017, www.census.gov/acs/www/data/data-tables-and-tools/data-profiles /2017.

25. "Advancing Sustainable Materials Management: 2017 Fact Sheet."

26. "Advancing Sustainable Materials Management: 2017 Fact Sheet"; William Rathje and Cullen Murphy, *Rubbish! The Archeology of Garbage* (Tucson: University of Arizona Press, 2001).

27. This does not mean that 25 percent of discards were actually recycled. Most plastics are never recycled, often because there is no end market for the quality of product that would result; many other materials that are put into recycling bins become too contaminated and are disposed of after they have been processed by a material recovery facility or recycling plant. Alex Cuyler, "Residual Reality with Single Stream Recycling," *BioCycle* 43, no. 6 (June 2002): 62; Eureka Recycling, "The Downstream of Single Stream," *Resource Recycling*, November 2002, www.eurekarecycling.org/pdfs

/ResourceRecyclingArticle.pdf; Laura Parker, "A Whopping 91 Percent of Plastic Isn't Recycled," *National Geographic News*, December 20, 2018, www.nationalgeographic.com /news/2017/07/plastic-produced-recycling-waste-ocean-trash-debris-environment.

28. Frank Ackerman, *Why Do We Recycle? Markets, Values, and Public Policy* (Washington, DC: Island Press, 1996); Samantha MacBride, *Recycling Reconsidered: The Present Failure and Future Promise of Environmental Action in the United States* (Cambridge, MA: MIT Press, 2012).

29. "Advancing Sustainable Materials Management: 2017 Fact Sheet."

30. US Congress, "Resource Conservation and Recovery Act," Pub. L. No. 94–580 (1976), www.govinfo.gov/content/pkg/STATUTE-90/pdf/STATUTE-90-Pg2795 .pdf.

31. "Disposal Bans and Mandatory Recycling in the United States," Northeast Recycling Council, July 2020, https://nerc.org/documents/disposal_bans_mandatory_re cycling_united_states.pdf; Marie Donahue, "Waste Incineration: A Dirty Secret in How States Define Renewable Energy," Institute for Local Self-Reliance, December 12, 2018, https://ilsr.org/wp-content/uploads/2018/12/ILSRIncinerationFInalDraft -6.pdf; "State Plastic and Paper Bag Legislation," National Conference of State Legislatures, January 24, 2020.

32. "Draft Massachusetts 2010–2020 Solid Waste Master Plan," Massachusetts Department of Environmental Protection, Executive Office of Energy and Environmental Affairs, July 1, 2010; "Draft Massachusetts 2030 Solid Waste Master Plan," Massachusetts Department of Environmental Protection, September 2019, www .mass.gov/guides/massdep-waste-disposal-bans#-waste-ban-regulations-&-poli cies-; "MassDEP Waste Disposal Bans," Massachusetts Department of Environmental Protection, 2020, www.mass.gov/guides/massdep-waste-disposal-bans#-waste-ban -regulations-&-policies-; "Commercial Organic Materials Waste Ban: Guidance for Businesses, Institutions, and Haulers," Massachusetts Department of Environmental Protection, June 2014, www.mass.gov/eea/docs/dep/recycle/laws/orgguid.pdf.

33. Samantha MacBride, "San Francisco's Famous 80 Percent Waste Diversion Rate: Anatomy of an Exemplar," *Discard Studies* (blog), December 6, 2013, http:// discardstudies.wordpress.com/2013/12/06/san-franciscos-famous-80-waste-diver sion-rate-anatomy-of-an-exemplar; MacBride, *Recycling Reconsidered*; Krones, "Accounting for Non-Hazardous Industrial Waste."

34. See, for example, Gregory C. Unruh, "Understanding Carbon Lock-In," *Energy Policy* 28, no. 12 (October 2000): 817–830, https://doi.org/10.1016/S0301-4215(00) 00070-7.

35. Horst W. J. Rittel and Melvin M. Webber, "Dilemmas in a General Theory of Planning," *Policy Sciences* 4 (1973): 161.

36. Brian W. Hogwood and Lewis A. Gunn, *Policy Analysis for the Real World* (Oxford: Oxford University Press, 1984); Robert Merton and Robert Nisbet, *Contemporary Social Problems: An Introduction to the Sociology of Deviant Behavior and Social Disorganization* (New York: Harcourt, Brace and World, 1961).

37. Deborah A. Stone, "Causal Stories and the Formation of Policy Agendas," *Political Science Quarterly* 104, no. 2 (Summer 1989): 299.

38. Joseph R. Gusfield, *The Culture of Public Problems: Drinking-Driving and the Symbolic Order* (Chicago: University of Chicago Press, 1981).

39. Lisa Bardwell, "Problem-Framing: A Perspective on Environmental Problem Framing," *Environmental Management* 15, no. 5 (1991): 603–612.

40. Stone, "Causal Stories," 281.

41. Gusfield, *Culture of Public Problems*; Stone, "Causal Stories"; Deborah Stone, *Policy Paradox: The Art of Political Decision-Making*, 3rd ed. (New York: W. W. Norton, 2012); John Kingdon, *Agendas, Alternatives, and Public Policies*, 2nd ed. (New York: Harper Collins College, 1995).

42. Jennifer S. Light, *From Warfare to Welfare: Defense Intellectuals and Urban Problems in Cold War America* (Baltimore: Johns Hopkins University Press, 2005).

43. Alan Black, "The Chicago Area Transportation Study: A Case Study of Rational Planning," *Journal of Planning Education and Research* 10, no. 1 (1990): 27–37; James C. Scott, *Seeing Like a State: How Certain Schemes to Improve the Human Condition Have Failed*, Yale Agrarian Studies (New Haven, CT: Yale University Press, 1998).

44. Martin V. Melosi, "Sanitary Engineers in American Cities: Changing Roles from the Age of Miasmas to the Age of Ecology," in *Civil Engineering History: Engineers Make History*, ed. Jerry R. Rogers, Donald Kennon, Robert T. Jaske, and Francis E. Griggs Jr. (New York: American Society of Civil Engineers, 1996); P. R. White, M. Franke, and P. Hindle, *Integrated Solid Waste Management: A Lifecycle Inventory* (Boston: Springer, 1995).

45. David Demeritt, "The Construction of Global Warming and the Politics of Science," *Annals of the Association of American Geographers* 91, no. 2 (June 2001): 307; Sheila Jasanoff, *Designs on Nature: Science and Democracy in Europe and the United States* (Princeton, NJ: Princeton University Press, 2005); Thomas S. Kuhn, *The Structure of Scientific Revolutions*, 2nd ed. (Chicago: University of Chicago Press, 1970).

46. Demeritt, "The Construction of Global Warming"; Jane Jacobs, *The Death and Life of Great American Cities* (New York: Random House, 1961); Charles E. Lindblom, "The Science of 'Muddling Through,'" *Public Administration Review* 19 (April 1959): 79–88.

47. Sheila Jasanoff, *States of Knowledge: The Co-Production of Science and the Social Order* (New York: Routledge, 2004); Naomi Oreskes and Erik M. Conway, *Merchants of Doubt: How a Handful of Scientists Obscured the Truth on Issues from Tobacco Smoke to Global Warming* (New York: Bloomsbury, 2010); Zahra Meghani and Jennifer Kuzma, "The 'Revolving Door' Between Regulatory Agencies and Industry: A Problem That Requires Reconceptualizing Objectivity," *Journal of Agricultural and Environmental Ethics* 24 (2011): 575–599, https://doi.org/10.1007/s10806-010-9287-x.

48. Stephen Graham and Simon Marvin, *Splintering Urbanism: Networked Infrastructures, Technological Mobilities and the Urban Condition* (New York: Routledge, 2001); Matthew Gandy, "Landscapes of Disaster: Water, Modernity, and Urban Fragmentation in Mumbai," *Environment and Planning A* 40, no. 1 (January 2008): 108–130, https://doi.org/10.1068/a3994.

49. Jurian Edelenbos, Arwin van Buuren, and Nienke van Schie, "Co-Producing Knowledge: Joint Knowledge Production Between Experts, Bureaucrats and Stakeholders in Dutch Water Management Projects," *Environmental Science and Policy* 14, no. 6 (October 2011): 675–684; Meghani and Kuzma, "The 'Revolving Door.'"

50. Clifford Geertz, *Local Knowledge: Further Essays in Interpretive Anthropology* (New York: Basic Books, 1983).

51. Charles Edward Lindblom and David K. Cohen, *Usable Knowledge: Social Science and Social Problem Solving* (New Haven, CT: Yale University Press, 1979).

52. Jonathan Murdoch and Judy Clark, "Sustainable Knowledge," *Geoforum* 25, no. 2 (May 1994): 115–132.

53. Alan Irwin, *Citizen Science: A Study of People, Expertise and Sustainable Development* (London: Routledge, 1995).

54. Brian Wynne, "Misunderstood Misunderstandings: Social Identities and Public Uptake of Science," in *Misunderstanding Science? The Public Reconstruction of Science and Technology*, ed. Brian Wynne and Alan Irwin (Cambridge: Cambridge University Press, 1996), 19–46.

55. Michael S. Carolan, "Science, Expertise and the Democratization of the Decision-Making Process," *Society and Natural Resources* 19, no. 7 (2008): 661–668; Sheila Jasanoff, "Public Knowledge, Private Fears," *Social Studies of Science* 27, no. 2 (1997): 350–355.

56. Jason Corburn, "Community Knowledge in Environmental Health Science: Co-Producing Policy Expertise," *Environmental Science and Policy* 10, no. 2 (April 2007): 150–161.

57. Anna Lehtonen, Arto O. Salonen, and Hannele Cantell, "Climate Change Education: A New Approach for a World of Wicked Problems," in *Sustainability, Human Well-Being, and the Future of Education*, ed. Justin W. Cook (Cham, Switzerland: Springer International, 2019), 339–374, https://doi.org/10.1007/978-3-319-78580-6_11.

58. Olivier Boiral, "Tacit Knowledge and Environmental Management," *Long Range Planning* 35 (2002): 291–317.

59. Matthew Harmin, M. J. Barrett, and Carolyn Hoessler, "Stretching the Boundaries of Transformative Sustainability Learning: On the Importance of Decolonizing Ways of Knowing and Relations with the More-than-Human," *Environmental Education Research* 23, no. 10 (November 2017): 1489–1500, https://doi.org/10.1080/135046 22.2016.1263279; Eve Tuck and K. Wayne Yang, "Decolonization Is Not a Metaphor," *Decolonization: Indigeneity, Education and Society* 1, no. 1 (2012): 1–40.

60. Jason Corburn, *Street Science: Community Knowledge and Environmental Health Justice*, Urban and Industrial Environments (Cambridge, MA: MIT Press, 2005); Brent Taylor and Rob C. de Loë, "Conceptualizations of Local Knowledge in Collaborative Environmental Governance," *Geoforum*, special issue, *Spatialities of Ageing* 43, no. 6 (November 2012): 1207–1217, https://doi.org/10.1016/j.geoforum.2012.03.007; Sylvia N. Tesh, "Citizen Experts in Environmental Risk," *Policy Sciences* 32 (1999): 39–58.

61. Taylor and de Loë, "Conceptualizations of Local Knowledge."

62. Jason Corburn, "Bringing Local Knowledge into Environmental Decision Making: Improving Urban Planning for Communities at Risk," *Journal of Planning Education and Research* 22, no. 4 (2003): 420–433.

63. Judith Petts and Catherine Brooks, "Expert Conceptualisations of the Role of Lay Knowledge in Environmental Decisionmaking: Challenges for Deliberative Democracy," *Environment and Planning A* 38, no. 6 (June 2006): 1047, https://doi.org/10.1068/a37373.

64. Harriet Bulkeley and Nicky Gregson, "Crossing the Threshold: Municipal Waste Policy and Household Waste Generation," *Environment and Planning A* 41,

no. 4 (April 2009): 929–945, https://doi.org/10.1068/a40261; Nicky Gregson, *Living with Things: Ridding, Accommodation, Dwelling* (Wantage, UK: Sean Kingston, 2011).

65. Max Liboiron, "Against Awareness, for Scale: Garbage Is Infrastructure, Not Behavior," *Discard Studies* (blog), January 23, 2014, https://discardstudies.com/2014/01/23/against-awareness-for-scale-garbage-is-infrastructure-not-behavior.

66. Archon Fung and Erik Olin Wright, *Deepening Democracy: Institutional Innovations in Empowered Participatory Governance* (London: Verso, 2003); Sherry R. Arnstein, "A Ladder of Citizen Participation," *Journal of the American Institute of Planners* 35, no. 4 (1969): 216–224.

67. Judith E. Innes, "Planning Through Consensus Building: A New View of the Comprehensive Planning Ideal," *Journal of the American Planning Association* 62, no. 4 (1996): 460–472, https://doi.org/10.1080/01944369608975712; Judith E. Innes and David E. Booher, *Planning with Complexity: An Introduction to Collaborative Rationality for Public Policy* (Abingdon, Oxon, UK: Routledge, 2010); Lawrence Susskind and Jeffrey L. Cruikshank, *Breaking the Impasse: Consensual Approaches to Resolving Public Disputes* (New York: Basic Books, 1987).

68. Judith Petts, "Enhancing Environmental Equity Through Decision-Making: Learning from Waste Management," *Local Environment* 10, no. 4 (August 1, 2005): 397–409, https://doi.org/10.1080/13549830500160933; Judith Petts, "Barriers to Participation and Deliberation in Risk Decisions: Evidence from Waste Management," *Journal of Risk Research* 7, no. 2 (March 2004): 115–133.

69. Innes and Booher, *Planning with Complexity*.

70. Wynne, "Misunderstood Misunderstandings."

71. Boiral, "Tacit Knowledge and Environmental Management"; Helena Mateus Jerónimo and José Luís Garcia, "Risks, Alternative Knowledge Strategies and Democratic Legitimacy: The Conflict over Co-Incineration of Hazardous Industrial Waste in Portugal," *Journal of Risk Research* 14, no. 8 (September 2011): 951–967, https://doi.org/10.1080/13669877.2011.571783.

72. Frank Fischer, *Citizens, Experts, and the Environment: The Politics of Local Knowledge* (Durham, NC: Duke University Press, 2000).

73. Michael K. Heinman, "Science by the People: Grassroots Environmental Monitoring and the Debate over Scientific Expertise," *Journal of Planning Education and Research* 16, no. 4 (1997): 291–199; Mirilia Bonnes, David Uzzell, Giuseppe Carrus, and Tanika Kelay, "Inhabitants' and Experts' Assessments of Environmental Quality for Urban Sustainability," *Journal of Social Issues* 63, no. 1 (March 2007): 59–78, https://doi.org/10.1111/j.1540-4560.2007.00496.x; Corburn, "Community Knowledge."

74. Taylor and de Loë, "Conceptualizations of Local Knowledge."

75. Fischer, *Citizens, Experts, and the Environment*; Judith Petts, "The Public-Expert Interface in Local Waste Management Decisions: Expertise, Credibility and Process," *Public Understanding of Science* 6, no. 4 (October 1997): 359–381, https://doi.org/10.1088/0963-6625/6/4/004.

76. Harmin et al., "Stretching the Boundaries"; Tuck and Yang, "Decolonization Is Not a Metaphor."

77. Arnstein, "Ladder of Citizen Participation."

78. Lindblom and Cohen, *Usable Knowledge*.

79. Michael S. Carolan, "Sustainable Agriculture, Science and the Co-Production of 'Expert' Knowledge: The Value of Interactional Expertise," *Local Environment* 11, no. 4 (January 2007): 422.

80. Camille Limoges, "Expert Knowledge and Decision-Making in Controversy Contexts," *Public Understanding of Science* 2, no. 4 (October 1993): 418, https://doi.org/10.1088/0963-6625/2/4/009.

81. Petts, "Public-Expert Interface."

82. Limoges, "Expert Knowledge"; Petts, "Public-Expert Interface"; Petts and Brooks, "Expert Conceptualisations."

83. Carolan, "Sustainable Agriculture"; Judith E. Innes, "Information in Communicative Planning," *Journal of the American Planning Association* 64, no. 1 (1998): 52–63.

84. John Friedmann, "Toward a Non-Euclidian Mode of Planning," *Journal of the American Planning Association* 59, no. 4 (September 1993): 482.

85. Antonella Maiello, Cláudia V. Viegas, Marco Frey, and José Luis D. Ribeiro, "Public Managers as Catalysts of Knowledge Co-Production? Investigating Knowledge Dynamics in Local Environmental Policy," *Environmental Science and Policy* 27 (March 2013): 141–150, https://doi.org/10.1016/j.envsci.2012.12.007.

86. Yvonne Rydin, "Re-Examining the Role of Knowledge Within Planning Theory," *Planning Theory* 6, no. 1 (March 2007): 52–68, https://doi.org/10.1177/1473095207075161.

87. "Zero Waste Case Study: Seattle," EPA, February 2019, www.epa.gov/transforming-waste-tool/zero-waste-case-study-seattle.

88. Scott Allen, "Boston Ranks 8th out of 13 US Cities on Recycling Rate," *Boston Globe*, July 30, 1993; "Boston About Results: Mayor's Quarterly Performance Report," City of Boston, Public Works and Transportation, January 1, 2011, www.cityofboston.gov/images_documents/PWD-BTD%20FY11%20Q3%20Web_ver1_tcm3-25820.pdf.

89. Hiroko Tabuchi, "In Coronavirus, Industry Sees Chance to Undo Plastic Bag Bans," *New York Times*, March 26, 2020, www.nytimes.com/2020/03/26/climate/plastic-bag-ban-virus.html; Ruth Brooke and Mark Sauer, "How the Coronavirus Has Impacted Plastic Pollution," *KPBS Public Media* (blog), July 2, 2020, www.kpbs.org/news/2020/jul/02/impact-coronavirus-plastic-pollution; Ashifa Kassam, "'More Masks Than Jellyfish': Coronavirus Waste Ends Up in Ocean," *Guardian*, June 8, 2020, www.theguardian.com/environment/2020/jun/08/more-masks-than-jellyfish-coronavirus-waste-ends-up-in-ocean; Rob Picheta, "Coronavirus Is Causing a Flurry of Plastic Waste," CNN, May 4, 2020, www.cnn.com/2020/05/04/world/coronavirus-plastic-waste-pollution-intl/index.html; Carol Konyn, "Another Side Effect of COVID-19: The Surge in Plastic Pollution," *Earth.org* (blog), July 6, 2020, https://earth.org/covid-19-surge-in-plastic-pollution.

Chapter 1. The Evolution of America's Weak Recycling Waste Regime

1. Martin V. Melosi, *The Sanitary City: Environmental Services in Urban America from Colonial Times to the Present*, abridged ed. (Pittsburgh: University of Pittsburgh Press, 2008).

2. Suellen M. Hoy, "'Municipal Housekeeping': The Role of Women in Improving Urban Sanitation Practices, 1880–1917," in *Pollution and Reform in American Cities, 1870–1930*, ed. Martin V. Melosi (Austin: University of Texas Press, 1980), 173–198; Suellen Hoy, *Chasing Dirt: The American Pursuit of Cleanliness* (repr., New York: Oxford University Press, 1996).

3. Jon A. Peterson, "The City Beautiful Movement: Forgotten Origins and Lost Meanings," *Journal of Urban History* 2, no. 4 (August 1976): 415–434, https://doi.org /10.1177/009614427600200402; Bonj Szczygiel, "'City Beautiful' Revisited: An Analysis of Nineteenth-Century Civic Improvement Efforts," *Journal of Urban History* 29, no. 2 (January 2003): 107–132, https://doi.org/10.1177/0096144202238870; Martin V. Melosi, *Garbage in the Cities: Refuse, Reform, and the Environment*, rev. ed. (Pittsburgh: University of Pittsburgh Press, 2005).

4. Melosi, *Garbage in the Cities*; Melosi, *Sanitary City*.

5. Melosi, *Garbage in the Cities*.

6. Stanley K. Schultz and Clay McShane, "To Engineer the Metropolis: Sewers, Sanitation, and City Planning in Late-Nineteenth-Century America," *Journal of American History* 65, no. 2 (September 1978): 399, https://doi.org/10.2307/1894086.

7. Martin V. Melosi, "Sanitary Engineers in American Cities: Changing Roles from the Age of Miasmas to the Age of Ecology," in *Civil Engineering History: Engineers Make History*, ed. Jerry R. Rogers, Donald Kennon, Robert T. Jaske, and Francis E. Griggs Jr. (New York: American Society of Civil Engineers, 1996), 111.

8. Melosi, *Garbage in the Cities*; Mansfield Merriman, *Elements of Sanitary Engineering* (New York: John Wiley, 1898); Benjamin Miller, *Fat of the Land: Garbage in New York, the Last Two Hundred Years* (New York: Four Walls Eight Windows, 2000).

9. William Francis Morse, *The Collection and Disposal of Municipal Waste* (New York: Municipal Journal and Engineer, 1908), 36.

10. Susan Strasser, *Waste and Want: A Social History of Trash* (New York: Metropolitan Books, 1999), 266.

11. Robin Nagle, "Garbage: Learning to Unsee" (Lecture, Trash Talk Lecture Series, Peabody Museum, Harvard University, September 14, 2011).

12. Franklin Delano Roosevelt, *Scrap Rubber Needed*, Universal Studios, June 15, 1942, Archive.org, https://archive.org/details/1942-06-15_Scrap_Rubber_Needed; *Your Scrap—Brought It Down—Keep Scrapping Iron and Steel-Rubber—All Other Metals-Rags—Move All Scrap Now!/Broder*, US Government Printing Office, 1942, www.loc .gov/item/90712783; Pennsylvania Art WPA, *Save Scrap for Victory! Save Metals, Save Paper, Save Rubber, Save Rags*, 1941–1943, www.loc.gov/resource/cph.3f05676.

13. Lizabeth Cohen, "A Consumers' Republic: The Politics of Mass Consumption in Postwar America," *Journal of Consumer Research* 31, no. 1 (June 2004): 236–239, https://doi.org/10.1086/383439.

14. Dolores Hayden, *Redesigning the American Dream: The Future of Housing, Work, and Family Life* (New York: W. W. Norton, 1984).

15. Lloyd Stouffer, "Plastics Packaging: Today and Tomorrow," Society of the Plastics Industry, November 1963, available at https://discardstudies.com/wp-content /uploads/2014/07/stoffer-plastics-packacing-today-and-tomorrow-1963.pdf; Cohen, "A Consumers' Republic."

16. "Throwaway Living," *Life*, August 1, 1955.

17. "Throwaway Living."

18. Arsen Darnay and William E. Franklin, "The Role of Packaging in Solid Waste Management, 1966–1976," Bureau of Solid Waste Management, 1969.

19. Irene Murray, "Garbage Problem," *Los Angeles Times*, June 26, 1969; A. Wright, "Garbage Along Colorado River," *Los Angeles Times*, July 5, 1969, B6.

20. Peter Franchot, "Bottles and Cans: The Story of the Vermont Deposit Law," National Wildlife Federation, 1978, http://vnrc.org/wp-content/uploads/2012/08/Bottes-and-Cans_the-story-of-the-vt-deposit-law.pdf.

21. Heather Rogers, *Gone Tomorrow: The Hidden Life of Garbage* (New York: New Press, 2005).

22. Thomas Barnes, "Framing Postwar Litter: The Photojournalism of Fenno Jacobs and Keep America Beautiful's Public Service Announcements," *Photography and Culture* 6, no. 2 (July 2013): 175–192, https://doi.org/10.2752/175145213X136068 38923273.

23. As quoted in Finis Dunaway, *Seeing Green: The Use and Abuse of Environmental Images* (Chicago: University of Chicago Press, 2015), 82.

24. Franchot, "Bottles and Cans"; Finn Arne Jørgensen, "A Pocket History of Bottle Recycling," *The Atlantic*, February 27, 2013, www.theatlantic.com/technology/archive/2013/02/a-pocket-history-of-bottle-recycling/273575.

25. Rachel Carson, *Silent Spring* (Boston: Houghton Mifflin, 1962).

26. Richard Corrigan, "City Dump Fouls Air While Officials Argue," *Washington Post*, March 7, 1966, B1; Homer Bigarts, "Great Swamp Endangered by Pollution from Town Dump," *New York Times*, November 18, 1968, 49; "Skokie to Air Debate over Garbage Unit," *Chicago Tribune*, November 8, 1962, n1; "Buried in Waste," *New York Times*, December 24, 1968, 22; "Chilling Prospect for Great Lakes," guest editorial, *Chicago Daily Tribune*, May 17, 1962, 22, reprinted from the *Detroit News*; S. Smith Griswold, "Smog Control Chief Defends Burning Ban," *Los Angeles Times*, November 14, 1964, B4; Charles S. Spoo, "Voice of the People: Air Pollution," *Chicago Daily Tribune*, April 21, 1962, 8.

27. Melosi, *Garbage in the Cities*, 201; Melosi, *Sanitary City*, 207–209.

28. Melosi, *Garbage in the Cities*; "The Solid Waste Dilemma: An Agenda for Action," Final Report of the Municipal Solid Waste Task Force, EPA, February 1989, http://nepis.epa.gov; "Methods to Manage and Control Plastic Wastes," EPA, February 1990, http://nepis.epa.gov.

29. Melosi, *Garbage in the Cities*, 168–189.

30. "The Solid Waste Dilemma: An Agenda for Action."

31. Martin V. Melosi, "The Cleaning of America," *Environment* 23, no. 8 (October 1981): 6; William Rathje and Cullen Murphy, *Rubbish! The Archeology of Garbage* (Tucson: University of Arizona Press, 2001); Strasser, *Waste and Want*.

32. And the trend has only continued. By 2017, plastics constituted approximately 13 percent of the national solid waste stream by weight; but since plastics are light, they likely constitute a much larger percentage by volume. "National Overview: Facts and Figures on Materials, Wastes and Recycling," EPA, March 2020, www.epa.gov/facts-and-figures-about-materials-waste-and-recycling/national-overview-facts-and-figures-materials.

33. *The Third Pollution*, Federal Solid Waste Management Program, Filmscripts on Solid Waste Management, EPA, 1966, National Service Center for Environmental

Publications, https://nepis.epa.gov/Exe/ZyPDF.cgi/910210VA.PDF?Dockey=910210 VA.PDF.

34. William E. Small, *Third Pollution: The National Problem of Solid Waste Disposal* (New York: Praeger, 1971), 4.

35. Small, *Third Pollution*, 100.

36. Gladwin Hill, "Major U.S. Cities Face Emergency in Trash Disposal: Growing National Problems May Parallel the Crisis in Air and Water Pollution," *New York Times*, June 16, 1969.

37. Peter Thorsheim, *Inventing Pollution: Coal, Smoke, and Culture in Britain Since 1800*, Series in Ecology and History (Athens: Ohio University Press, 2006).

38. Mary Douglas, *Purity and Danger: An Analysis of the Concepts of Pollution and Taboo*, new ed. (New York: Routledge, 1984), 2.

39. Carson, *Silent Spring*; Theo Colborn, Dianne Dumanoski, and John Peterson Myers, *Our Stolen Future: Are We Threatening Our Fertility, Intelligence, and Survival? A Scientific Detective Story* (New York: Dutton, 1996); Devra Lee Davis, *When Smoke Ran Like Water: Tales of Environmental Deception and the Battle Against Pollution* (New York: Basic Books, 2002); Naomi Oreskes and Erik M. Conway, *Merchants of Doubt: How a Handful of Scientists Obscured the Truth on Issues from Tobacco Smoke to Global Warming* (New York: Bloomsbury, 2010); Amy Davidson, "The Contempt That Poisoned Flint's Water," *New Yorker*, January 22, 2016, www.newyorker.com/news/amy -davidson/the-contempt-that-poisoned-flints-water.

40. Douglas, *Purity and Danger*, 2.

41. Nina Wormbs, "Radio Pollution: From Sparks to White Spots," conference paper, American Society for Environmental History, Seattle, April 1, 2016.

42. William Cronon, *Changes in the Land: Indians, Colonists, and the Ecology of New England* (New York: Farrar, Straus and Giroux, 1983); William Cronon, *Nature's Metropolis* (New York: W. W. Norton, 1992); David E. Nye, *American Technological Sublime* (Cambridge, MA: MIT Press, 1994); Anne Whiston Spirn, *The Language of Landscape* (New Haven, CT: Yale University Press, 1998); Leo Marx, *The Machine in the Garden: Technology and the Pastoral Ideal in America* (New York: Oxford University Press, 2000); Roderick Nash, *Wilderness and the American Mind* (New Haven, CT: Yale University Press, 2001); Ted Steinberg, *Down to Earth: Nature's Role in American History* (New York: Oxford University Press, 2002).

43. Philip S. Gutis, "Beach Waste Raises New Fears," *New York Times*, July 17, 1988, LI1; Peter Crescenti, "Will the Beaches Be Clean and Crowded?," *New York Times*, May 28, 1989, LI1; Eric Schmitt, "Summer Cranks Up, Anti-Trash Armor Hits the Beach," *New York Times*, May 23, 1989, A1; Joseph F. Sullivan, "Less Debris on Beaches Is Predicted," *New York Times*, February 25, 1989, Metro 31.

44. Martin Waldron, "Millions Are Spent to Fight Pollution of Ocean Beaches," *New York Times*, June 29, 1970, 1; Gordon Grant, "Sewage Keeps Beaches Closed," *Los Angeles Times*, March 29, 1978, OC1; James Quinn, "Raw Sewage Closes Beaches for Week," *Los Angeles Times*, February 21, 1980, V1; Marjorie Miller, "Sewage Closes Imperial Beach Oceanfront," *Los Angeles Times*, December 28, 1984, A1; Jane Gross, "Beaches on S.I. and Brooklyn Close as Sewage Hits Harbor," *New York Times*, July 13, 1988, A1.

45. Thomas Burton and George Papajohn, "Landfill Laws Fail to Do Job," *Chicago Tribune*, December 22, 1987, A1; Mike Ward, "Montebello Fears Impact of Pos-

sible Toxic Cleanup from Nearby Dump," *Los Angeles Times*, August 3, 1986, SE1; Irene Chang, "EPA Plans More Test Wells Near Toxic Landfill: Pollution," *Los Angeles Times*, November 10, 1991, SGV1; Suzanne DeChillo, "Fishermen's Group Keeps Pressure on County over Dump," *New York Times*, August 3, 1986, WC1.

46. Pete Earley, "EPA Says Garbage-Burning May Produce Toxic Chemical," *Boston Globe*, November 24, 1983, A3; Pete Earley, "Dioxin from Garbage Incinerators May Be Polluting Air Around Cities," *Washington Post*, October 20, 1983, A21; Stevenson Swanson, "City's Incinerator Must Clean Up Act," *Chicago Tribune*, December 5, 1993, NWA1; Andrew Pollack, "In Japan's Burnt Trash, Dioxin Threat: Near Incinerators, a Study Finds an Ominously High Infant Death Rate," *New York Times*, April 27, 1997, International, 10.

47. Small, *Third Pollution*, 7.

48. Maria Kaika, *City of Flows: Modernity, Nature and the City* (New York: Routledge, 2005).

49. Richard Rothstein, *The Color of Law: A Forgotten History of How Our Government Segregated America* (repr., Liveright, 2017); Peter Dreier, John Mollenkopf, and Todd Swanstrom, *Place Matters: Metropolitics for the Twenty-First Century*, 3rd ed. (Lawrence: University Press of Kansas, 2014); Carl A. Zimring, *Clean and White: A History of Environmental Racism in the United States* (New York: NYU Press, 2016).

50. Melvin Whitaker, "Yes, Blacks Oppose Pollution," *Sacramento Observer*, May 4, 1970.

51. Zimring, *Clean and White*; Dorceta E. Taylor, "The Rise of the Environmental Justice Paradigm Injustice Framing and the Social Construction of Environmental Discourses," *American Behavioral Scientist* 43, no. 4 (January 2000): 508–580, https://doi.org/10.1177/0002764200043004003; Robert D. Bullard, *Dumping in Dixie: Race, Class and Environmental Quality* (Boulder: Westview Press, 1994).

52. Luke W. Cole and Sheila R. Foster, *From the Ground Up: Environmental Racism and the Rise of the Environmental Justice Movement* (New York: NYU Press, 2001).

53. William May, as quoted in Small, *Third Pollution*, 99.

54. As quoted in Adam Rome, *The Genius of Earth Day: How a 1970 Teach-In Unexpectedly Made the First Green Generation* (New York: Hill and Wang, 2013), 86.

55. Rome, *The Genius of Earth Day*.

56. David Bird, "In the Aftermath of Earth Day: City Gains New Leverage," *New York Times*, April 24, 1970, 28.

57. Stefan Kanfer, "Going Pogo: The Life and Times of Walt Kelly's Political Possum," *City Journal*, Autumn 2011, www.city-journal.org/html/going-pogo-13429.html; Walt Kelly, *We Have Met the Enemy and He Is Us* (New York: Simon and Schuster, 1972).

58. "New Law to Control Hazardous Wastes, End Open Dumping, Promote Conservation of Resources," EPA, December 13, 1976, www.epa.gov/aboutepa/new-law-control-hazardous-wastes-end-open-dumping-promote-conservation-resources.

59. Garrick E. Louis, "A Historical Context of Municipal Solid Waste Management in the United States," *Waste Management and Research* 22 (2004): 317.

60. "The Waste System," EPA, Report No. 000K8801, November 1988.

61. Louis, "Historical Context."

62. Louis Blumberg and Robert Gottlieb, *War on Waste: Can America Win Its Battle*

with Garbage? (Covelo, CA: Island Press, 1989), 4; Melosi, *Sanitary City*, 240; Rogers, *Gone Tomorrow*, 200.

63. Josh Barbanel, "Garbage Crisis: After Landfills, What? Counties Ponder Disposal Problem," *New York Times*, April 2, 1978, NJ1; Bruce Keppel, "Dump Closes Down as 'Garbage Crisis' Looms," *Los Angeles Times*, January 2, 1981, C1; "The Coming Garbage Crisis," *Chicago Tribune*, January 11, 1986, A10; Edward Hudson, "Garbage Crisis: Landfills Are Nearly Out of Space," *New York Times*, April 4, 1986, B2; Joshua Gordon and Takoma Park, "The Municipal Garbage Crisis," *Washington Post*, December 24, 1987, A14; Iver Peterson, "Mounds of Garbage Signal Landfill Crisis in Jersey," *New York Times*, April 16, 1987, B1; Philip Shabecoff, "With No Room at the Dump, U.S. Faces a Garbage Crisis," *New York Times*, June 29, 1987, B8; Peg McDonnell Breslin, "Garbage Crisis," *Chicago Tribune*, January 31, 1988, C2; Don Oldenburg, "Garbage! America Faces a Growing Crisis of Volume," *Los Angeles Times*, October 16, 1988, 3; Martin Zimmerman, "Garbage Crisis Offers No Easy Solutions," *Los Angeles Times*, November 1, 1989, F12.

64. Patricia Poore, "America's 'Garbage Crisis': A Toxic Myth," *Harper's* 288, no. 1726 (March 1994): 24; Martin V. Melosi, "The 'Garbage Crisis' and the Weight of History," *Journal of Urban Technology* 1, no. 3 (1994): 1–20; William L. Rathje, "Rubbish!," *The Atlantic*, December 1989.

65. Harold Crooks, *Dirty Business: The Inside Story of the New Garbage Agglomerates* (Toronto: James Lorimer and Company, 1983); Harold Crooks, *Giants of Garbage* (Toronto: James Lorimer and Company, 1993); Garrick E. Louis, "A Historical Context of Municipal Solid Waste Management in the United States," *Waste Management and Research* 22, no. 4 (August 2004): 306–322, https://doi.org/10.1177/0734242X04 045425.

66. Melosi, *Sanitary City*, 241.

67. Henry Schwartz, Congressional Testimony, United States Congress House Committee on Science and Astronautics, "Technology Assessment—1970: Hearings Before the Subcommittee on Science, Research, and Development of the Committee on Science and Astronautics," 91st Cong., 2nd sess. (H.R. 17046, US Government Printing Office, 1970), 880; Barry Commoner, *The Closing Circle: Nature, Man and Technology* (New York: Random House, 1971); Small, *Third Pollution*; Susan Hulsman Bingham and Sherry Koehler, *It's Your Environment: Things to Talk About, Things to Do*, Environmental Action Coalition (New York: Scribner, 1976).

68. EPA, "Effective Hazardous Waste Management (Non-Radioactive); Position Statement," *Federal Register* 41, no. 161 (August 18, 1976): 35049–35051.

69. "The Solid Waste Dilemma: An Agenda for Action."

70. "The Solid Waste Dilemma: An Agenda for Action," 20.

71. Bartow J. Elmore, *Citizen Coke: The Making of Coca-Cola Capitalism* (New York: W. W. Norton, 2016), 242.

72. George J. Church, Steven Holmes, and Elizabeth Taylor, "Garbage, Garbage, Everywhere: Landfills Are Overflowing, but Alternatives Are Few," *Time* 132, no. 10 (September 5, 1988): 81; Hudson, "Garbage Crisis"; Keppel, "Garbage Crisis."

73. "Solid Waste Management and Resource Recovery, Technical Assistance Handbook," State of Florida, 1976; "Community Exchange Series: Resource Recovery," Metropolitan Area Planning Council, 1980; "The Solid Waste Dilemma: An

Agenda for Action"; Marie Cocco, "Locals Left Holding the Bag," in *Rush to Burn: Solving America's Garbage Crisis* (Washington, DC: Island Press, 1989 [series first published in *Newsday*, 1987]), 127–139; Richard C. Firstman, "High Stakes Risk on Incinerators," in *Rush to Burn*, 8; Shabecoff, "No Room at the Dump."

74. Owen J. Furuseth and Janet O'Callaghan, "Community Response to a Municipal Waste Incinerator: NIMBY or Neighbor?," *Landscape and Urban Planning* 21, no. 3 (November 1991): 163–171, https://doi.org/10.1016/0169-2046(91)90015-E; Thomas H. Rasmussen, "Not in My Backyard: The Politics of Siting Prisons, Landfills, and Incinerators," *State and Local Government Review* 24, no. 3 (1992): 128–134; Merrie Klapp, "Bargaining with Uncertainty: The Brooklyn Navy Yard Incinerator Dispute," *Journal of Planning Education and Research* 8, no. 3 (July 1989): 157–166, https://doi.org/10.1177/0739456X8900800303; Melissa Checker, "'Like Nixon Coming to China': Finding Common Ground in a Multi-Ethnic Coalition for Environmental Justice," *Anthropological Quarterly* 74, no. 3 (2001): 135–146.

75. "Waste Incinerators: Bad News for Recycling," Global Alliance for Incinerator Alternatives (GAIA hereafter), October 2013, www.no-burn.org; "Burning Public Money for Dirty Energy: Misdirected Subsidies for 'Waste-to-Energy' Incinerators," GAIA, November 2011; Lucie Laurian and Richard Funderburg, "Environmental Justice in France? A Spatio-Temporal Analysis of Incinerator Location," *Journal of Environmental Planning and Management* 57, no. 3 (March 2014): 424–446, https://doi.org/10.1080/09640568.2012.749395.

76. Massachusetts actually lifted the moratorium in 2013, after twenty-three years, though no new incinerators have been built. Jo Chesto, "Gov. Deval Patrick Lifts 23-Year-Old Moratorium on Incinerators in Massachusetts," *Boston Business Journal*, May 7, 2013, www.bizjournals.com/boston/blog/mass_roundup/2013/05/deval-patrick-lifts-incinerator-ban.html.

77. George Tchobanoglous, Hilary Theisen, and Samuel Vigil, *Integrated Solid Waste Management: Engineering Principles and Management Issues* (New York: McGraw-Hill, 1993).

78. P. R. White, M. Franke, and P. Hindle, *Integrated Solid Waste Management: A Lifecycle Inventory* (Boston: Springer, 1995), 1.

79. Mushtaq A. Memon, "Integrated Solid Waste Management Based on the 3R Approach," *Journal of Material Cycles and Waste Management* 12, no. 1 (2010): 30–40.

80. Memon, "Integrated Solid Waste Management Based on the 3R Approach"; "What Is Integrated Solid Waste Management?," EPA, May 2002, www.epa.gov/climatechange/wycd/waste/downloads/overview.pdf; White et al., *Integrated Solid Waste Management*, 11.

81. "Waste Incinerators: Bad News for Recycling"; "Trash Incineration More Polluting Than Coal," Energy Justice Network, n.d., www.energyjustice.net/incineration/worsethancoal; "DC Council: Reject the Covanta Waste Contract," Energy Justice Network, 2015, www.energyjustice.net/dc/wastecontract.

82. White et al., *Integrated Solid Waste Management*, 14.

83. "Advancing Sustainable Materials Management: 2017 Fact Sheet," EPA, Office of Land and Emergency Management, November 2019, www.epa.gov/sites/production/files/2019-11/documents/2017_facts_and_figures_fact_sheet_final.pdf; "Recycling Works: State and Local Solutions to Solid Waste Management Problems," EPA, Office of Solid Waste, January 1989, available at https://nepis.epa.gov.

84. Schwartz, Congressional Testimony, 880; see also T. Randall Curlee, *The Economic Feasibility of Recycling: A Case Study of Plastic Wastes* (Santa Barbara, CA: Praeger, 1986).

85. Matthew Gandy, *Recycling and the Politics of Urban Waste* (London: Earthscan, 1994); "Massachusetts Container Deposit Return System," Container Recycling Institute, 2017, available at www.container-recycling.org/index.php/publications/cri -publications; Jenny Gitlitz, "Bottled Up: Beverage Container Recycling Stagnates (2000–2010)," Container Recycling Institute, 2013; Finn Arne Jørgensen, "A Pocket History of Bottle Recycling," *The Atlantic*, February 27, 2013, www.theatlantic.com /technology/archive/2013/02/a-pocket-history-of-bottle-recycling/273575.

86. Bartow J. Elmore, "The American Beverage Industry and the Development of Curbside Recycling Programs, 1950–2000," *Business History Review* 86, no. 3 (2012): 477–501.

87. "Plastics: Material-Specific Data," Collections and Lists, EPA, September 12, 2017, www.epa.gov/facts-and-figures-about-materials-waste-and-recycling/plastics -material-specific-data; Ivars Peterson, "New Life for Old Plastics," *Science News* 126, no. 9 (1984): 140–141, https://doi.org/10.2307/3969096; J. R. Lawrence, "Status Report on Plastics Recycling," *Disposal of Plastics with Minimum Environmental Impact*, January 1973, https://doi.org/10.1520/STP38585S.

88. Samantha MacBride, *Recycling Reconsidered: The Present Failure and Future Promise of Environmental Action in the United States* (Cambridge, MA: MIT Press, 2012).

89. "Recycling Works."

90. MacBride, *Recycling Reconsidered*; Elmore, "American Beverage Industry"; Elmore, *Citizen Coke*.

91. Blumberg and Gottlieb, *War on Waste*; Adam S. Weinberg, David N. Pellow, and Allan Schnaiberg, *Urban Recycling and the Search for Sustainable Community Development* (Princeton, NJ: Princeton University Press, 2000).

92. Timothy W. Luke, *Ecocritique: Contesting the Politics of Nature, Economy, and Culture* (Minneapolis: University of Minnesota Press, 1997); MacBride, *Recycling Reconsidered*; Frank Ackerman, *Why Do We Recycle? Markets, Values, and Public Policy* (Washington, DC: Island Press, 1996).

93. Gay Hawkins, "Myths of the Circular Economy," *Discard Studies* (blog), November 18, 2019, https://discardstudies.com/2019/11/18/myths-of-the-circular-economy.

94. Zsuzsa Gille, *From the Cult of Waste to the Trash Heap of History: The Politics of Waste in Socialist and Postsocialist Hungary* (Bloomington: Indiana University Press, 2007).

95. Elizabeth Royte, *Garbage Land* (New York: Back Bay Books/Little, Brown, 2005); Annie Leonard, *The Story of Stuff*, Free Range Studios, 2007; Nagle, "Garbage: Learning to Unsee"; Mira Engler, *Designing America's Waste Landscapes* (Baltimore: Johns Hopkins University Press, 2004).

Chapter 2. Non-Planning for Garbage in Boston

1. Karl Haglund, *Inventing the Charles River* (Cambridge, MA: MIT Press, 2003); Nancy S. Seasholes, *Gaining Ground: A History of Landmaking in Boston* (Cambridge, MA: MIT Press, 2003).

2. Doug Most, *The Race Underground: Boston, New York, and the Incredible Rivalry That Built America's First Subway* (New York: St. Martin's, 2014); Joe McKendry, *Beneath the Streets of Boston: Building America's First Subway* (Boston: David R. Godine, 2005).

3. Thomas H. O'Connor, *Building a New Boston: Politics and Urban Renewal, 1950–1970* (Boston: Northeastern University Press, 1993); Lizabeth Cohen, "Buying into Downtown Revival: The Centrality of Retail to Postwar Urban Renewal in American Cities," *Annals of the American Academy of Political and Social Science* 611 (2007): 82–95.

4. Michael R. Fein, "Tunnel Vision: 'Invisible' Highways and Boston's 'Big Dig' in the Age of Privatization," *Journal of Planning History* 11, no. 1 (February 2012): 47–69, https://doi.org/10.1177/1538513211425209; F. P. Salvucci, "The 'Big Dig' of Boston, Massachusetts: Lessons to Learn," in *(Re)claiming the Underground Space: Proceedings of the ITA World Tunnelling Congress 2003*, ed. Jan Saveur (Amsterdam: CRC Press, 2003).

5. Thomas Andrew Savage, "Boston Harbor: The Anatomy of a Court-Run Cleanup," *Boston College Environmental Affairs Law Review* 22, no. 2 (December 1995); Eric Jay Dolin, *Political Waters: The Long, Dirty, Contentious, Incredibly Expensive, but Eventually Triumphant History of Boston Harbor. A Unique Environmental Success Story* (Amherst: University of Massachusetts Press, 2004).

6. Herbert J. Gans, "The Failure of Urban Renewal," *Commentary*, April 1965, www.commentarymagazine.com/articles/the-failure-of-urban-renewal; Herbert J. Gans, *Urban Villagers: Group and Class in the Life of Italian-Americans* (New York: London: Free Press, 1982).

7. Charles W. Millard, "The New Boston: City Hall," *Hudson Review* 23, no. 1 (1970): 110–115, https://doi.org/10.2307/3849613; Mark Pasnik, Michael Kubo, and Chris Grimley, *Heroic: Concrete Architecture and the New Boston* (New York: Monacelli Press, 2015).

8. Kayo Tajima, "New Estimates of the Demand for Urban Green Space: Implications for Valuing the Environmental Benefits of Boston's Big Dig Project," *Journal of Urban Affairs* 25, no. 5 (December 2003): 641–655, https://doi.org/10.1111/j.1467-9906 .2003.00006.x.

9. Susan Leigh Star, "Ethnography of Infrastructure," *American Behavioral Scientist* 43, no. 3 (December 1999): 90.

10. Martin V. Melosi, *The Sanitary City: Environmental Services in Urban America from Colonial Times to the Present*, abridged ed. (Pittsburgh: University of Pittsburgh Press, 2008).

11. Rudolph Hering and Samuel Greeley, *Collection and Disposal of Municipal Refuse* (New York: McGraw-Hill, 1921), 90.

12. "Hallet Street Dump Study," Boston Redevelopment Authority, 1970, BPL Gov Docs 143 July; "Columbia Point: Composite Development Plan," Columbia Point Peninsula Planning Committee, 1977, BPL Gov Docs BRA/105.

13. Harold Crooks, *Giants of Garbage* (Toronto: James Lorimer and Company, 1993); Harold Crooks, *Dirty Business: The Inside Story of the New Garbage Agglomerates* (Toronto: James Lorimer and Company, 1983); *Annual Report*, Boston Public Works Department, 1976–1977, and *Annual Report*, Boston Public Works Department, 1979–1980, Boston Municipal Archives (BMA hereafter), PWD Annual Reports 1966–1982.

14. Wendy Fox, "Problem: Finding the Best Way to Throw It Away," *Boston Globe*, May 28, 1983, B1; Mayor's Office of Energy Conservation, *The Incinerator at South Bay, Its History and Future* (Boston, 1978).

15. *Annual Report*, Boston Public Works Department, 1976, BMA/DWP Annual Reports 1966–1982; US Congress, Office of Technology Assessment, *Materials and Energy from Municipal Waste* (Washington, DC: US Government Printing Office, 1979); Gordian Associates, "Overcoming Institutionalized Barriers to Solid Waste Utilization as an Energy Source," US Department of Energy, Division of Synthetic Fuels, November 1977.

16. Fox, "Problem: Finding the Best Way to Throw It Away."

17. Peter Sleeper, "Waste Dilemma: Quincy Typifies a Wider Problem," *Boston Globe*, May 3, 1985; Andrew J. Dabilis, "Dukakis in Crisis on Landfills," *Boston Globe*, May 5, 1985; "Trash Haulers to Circle Common Today in Bid for More Landfills," *Boston Globe*, July 10, 1984.

18. Zsuzsa Gille, *From the Cult of Waste to the Trash Heap of History: The Politics of Waste in Socialist and Postsocialist Hungary* (Bloomington: Indiana University Press, 2007).

19. Jerry Ackerman, "Landfills: A Cure That Brought New Ills," *Boston Globe*, May 2, 1985.

20. Chris Chinlund and Peter Sleeper, "Eastern Mass. Landfill Crisis: The Symptoms Are Growing," *Boston Globe*, May 13, 1984, 1; Jerry Ackerman, "Threats to Water Purity Mounting," *Boston Globe*, October 15, 1985; Jerry Ackerman, "Homeowners Are Dumping Toxic Wastes," *Boston Globe*, November 12, 1984, 1; John Milne, "Questions of Safety Raised over Leaking N.E. Landfills," *Boston Globe*, May 11, 1986, A50.

21. "The Mayor's FY 1983 Budget Recommendations," Boston Municipal Research Bureau, July 16, 1982, https://archive.org/stream/mayorsfy1983budg00bost#page/n0/mode/2up.

22. Letter from Keep American Beautiful president Roger Powers to Joseph Casazza, December 2, 1981, BMA/Casazza Files/Box 27/Keep America Beautiful.

23. Memorandum from Frederick Betzner to Joe Casazza, Re: Modification of Trash Ordinance, July 30, 1979, BMA/Casazza Files/Box 23—1980/Rubbish Memorandum from Deputy Mayor Katherine Kane to Joseph Casazza, Re: Enclosed Article, July 28, 1980; memorandum from Joseph Sances to Joe Casazza, Re: Cleveland Circle Area Litter Baskets, November 14, 1980; letter from Joe Casazza to Roberta Kelman, executive director of the Beacon Hill Civic Association, August 21, 1980; letter from Joe Casazza to Muriel Davis, August 21, 1980; letter from Joe Casazza to Paul D'Addario, July 3, 1980; memorandum from Joseph Sanees to Brian White, Re: Litter Baskets, July 3, 1980, BMA/Casazza Files/Box 23—1980/Sanitary Division. Memorandum from Cornelius Doherty to Joe Casazza, Re: Survey of Litter Baskets, August 21, 1981, BMA/Casazza Files/Box 27—1982/Litter Baskets. Memorandum from Arnold Spector to Joe Casazza, Re: A Litter Enforcement Program, May 19,

1982, with attachments; memorandum from Eugenie Beal to Joe Casazza, Re: Update on Litter Enforcement Program, June 14, 1982, BMA/Casazza Files/Box 27/Environment Department. Memorandum from Eugenie Beal to Alex Taft, Re: Cleanup Recommendations, July 29, 1982; City of Boston Environment Department, Litter Enforcement Project Summary, undated, BMA/Casazza Files/Box 27/Environment Department.

24. Letter from Joe Casazza to Paul D'Addario, July 3, 1980; letter from Paul D'Addario to Joe Casazza, June 10, 1980, BMA/Casazza Files/Box 23—1980/Sanitary Division.

25. Memorandum from Joseph Sances to Joe Casazza, Re: Cleveland Circle Area Litter Baskets, November 14, 1980; letter from Joe Casazza to Roberta Kelman, executive director of the Beacon Hill Civic Association, August 21, 1980; memorandum from Joseph Sanees to Brian White, Re: Litter Baskets, July 3, 1980, BMA/Casazza Files/Box 23—1980/Sanitary Division.

26. Ian Menzies, "In Pursuit of Cleanliness," *Boston Globe*, November 3, 1983, B1.

27. Fox, "Problem: Finding the Best Way to Throw It Away"; M. E. Malone, "Boston's Fixture in Public Works," *Boston Globe*, February 29, 1988, B2.

28. In further evidence of industry consolidation, BFI is now a subsidiary of Waste Management, Inc.

29. Ed Quill, "Flynn Asks What the Rush Is on OK for Waste-to-Energy Plant," *Boston Globe*, December 14, 1983, Metro 37.

30. Jerry Ackerman, "Environment: Making Energy from Our Trash," *Boston Globe*, June 11, 1984, C1; Fox, "Problem: Finding the Best Way to Throw It Away."

31. Quill, "Flynn Asks What the Rush Is."

32. "Trash Haulers to Circle Common Today."

33. Peter Sleeper, "Dukakis to File Bill to Smooth Way for Landfills," *Boston Globe*, July 11, 1984, 18. Later in 1984, the state began to articulate a more comprehensive policy, but it ultimately did not follow through on any of the legislative proposals made immediately after the protest.

34. "The Politics of Waste Disposal," editorial, *Boston Globe*, May 3, 1985, 18.

35. Sleeper, "Dukakis to File Bill."

36. James Simon, "State Still in Quandary over Garbage," *Boston Globe*, April 22, 1985, Metro 17.

37. Letter from Richard Chaplin, DEQE, to Clifford Jessberger, American REF-Fuel, January 18, 1985, BMA/Casazza Files/Box 32/WTE/Folder—American REF-Fuel, Inc.

38. Ed Quill, "Officials Say Plan to Take Solid Waste out of City Costs More," *Boston Globe*, July 25, 1985, Metro 19.

39. Peter Sleeper, "Flynn, Dukakis Reach No Accord on Trash Disposal, Money at Issue," *Boston Globe*, December 24, 1987, B1.

40. Peter Howe, "Flynn Derides Dukakis Role in Hub Trash," *Boston Globe*, December 22, 1988.

41. Larry Tye, "Coalition Advocates Recycling as Alternative to Incinerators," *Boston Globe*, August 7, 1987, Metro 15.

42. Larry Tye, "Trash Plant Is Good Neighbor, Many Say," *Boston Globe*, May 20, 1986, B1.

43. Tye, "Coalition Advocates Recycling." It is important to note here that most

environmental advocates view incineration as antithetical to recycling. The most efficient combustion requires materials like plastic and paper that could otherwise be recycled. It also requires a consistent—and large—amount of waste material. To ensure
this feedstock, incinerators often include "put-or-pay" clauses in contracts, forcing
municipalities to supply waste or pay extra fees; this requirement discourages municipalities from reducing their waste streams through recycling or other means. In other
words, of course incinerator advocates would discourage recycling; it threatens their
business.

44. Larry Tye, "Mass. Dilemma: Taking Out the Trash," *Boston Globe*, June 9, 1988,
Metro 29.

45. Neil Sullivan, personal interview, May 9, 2016.

46. Ed Quill, "Boston to Truck Trash out of the City," *Boston Globe*, August 3,
1985, B1.

47. Jennifer Gabrys, "The Distancing of Waste: Overconsumption in a Global
Economy," in *Confronting Consumption*, ed. Thomas Princen, Michael Maniates, and
Ken Conca (Cambridge, MA: MIT Press, 2002), 155–176.

48. CSI Resource Systems, Inc., "Draft Environmental Impact Report: City of
Boston Waste-to-Energy Project," City of Boston, Department of Public Works,
December 15, 1983, 1, Boston Public Library (BPL hereafter)/GOV DOCS M3.B16/
PW/83.1/VOL. 1.

49. Citizen Advisory Committee on Solid Waste Disposal, "Report to Mayor Raymond L. Flynn: Issues to Consider in Making a Decision on Long-Term Solid Waste
Disposal," July 29, 1985, BMA/Collection 0246/Box V-9/Waste-to-Energy Incinerator 2-2.

50. Citizen Advisory Committee, "Issues to Consider."

51. Citizen Advisory Committee, "Issues to Consider"; M. E. Malone, "Advisory
Panel on Disposal Meets Tonight," *Boston Globe*, August 11, 1987, Metro 19; M. E.
Malone, "Flynn Refuses to Ask Group to Look at Other Waste Sites," *Boston Globe*,
August 12, 1987, Metro 42.

52. Bruce Mohl and M. E. Malone, "Bulger Looks Outside Boston for Trash-
Disposal Aid," *Boston Globe*, July 24, 1987, Metro 13.

53. Peggy Hernandez, "Council Hopefuls Air Trash Proposals: Candidates Agree
Waste Must Be Recycled," *Boston Globe*, October 8, 1987, Metro 93; "Incinerator Insight," editorial, *Boston Globe*, September 8, 1987, A14.

54. Excerpt from Meeting of Special Committee on Waste Management, December 7, 1988 (2), BMA/0246.001/Box V-9/4-4.

55. Ordinance Establishing a Solid Waste Management System for the City of
Boston, Summary, Boston City Council, BMA/0246.001/Box V-9/4-4; Special Committee on Solid Waste Management, "Dealing with Our Trash: A Report to the Boston City Council," October 3, 1988.

56. Peter Howe, "Recycling Stressed in Councilors' Trash Plan," *Boston Globe*,
December 9, 1988, Metro 25.

57. Peter Howe, "Flynn, Scondras Disagree over Need for Fees, Fines in Recycling
Plan," *Boston Globe*, December 12, 1988, Metro 19.

58. In response to the evident public interest in non-disposal alternatives surfaced
by the city council's report, Mayor Flynn proposed a minimal recycling ordinance that
required residential recycling but contained no enforcement mechanisms and made no

commitments for the provision of infrastructure or service from the city. The recycling plan will be discussed in detail in the following chapter.

59. "Coping with the Solid Waste Crisis: A Practical Guide for Local Officials and Citizens," Metropolitan Area Planning Council, Boston, January 1986.

60. "Getting Boston's Solid Waste out of the Dumps," Boston Municipal Research Bureau, November 17, 1982, BPL/GOV DOCS BMR/S3/82-9 (R).

61. "Financial Comparison of Boston's Solid Waste Disposal Options," Boston Municipal Research Bureau, February 1986, on file with the BMRB, 36–38.

62. "Financial Comparison of Boston's Solid Waste Disposal Options."

63. Quill, "Officials Say Plan"; Ed Quill, "Trash Bills for Boston May Double," *Boston Globe*, April 8, 1986, B1; Ed Quill, "Taking Trash to Suburbs Seen as Costing City $215M," *Boston Globe*, February 20, 1986, Metro 27; Quill, "Boston to Truck Trash."

64. Ed Quill, "Trash Plan Meeting Opposition," *Boston Globe*, May 4, 1986, Metro 29; Ed Quill, "Flynn's Switch on Trash Plant Site Fuels Debate over Decision Making," *Boston Globe*, May 11, 1986, Metro 31; Carol Pearson, "Flynn Vows to Develop Area Slated for Jail, Trash Plant," *Boston Globe*, May 16, 1986, Metro 75; Alexander Reid, "South Bay Trash Incinerator Nearing Gantlet of Hearings," *Boston Globe*, June 28, 1987, Metro 98; Tye, "Coalition Advocates Recycling"; Michael K. Frisby, "Candidate Forum Seeks Answers on Tenants, Trash," *Boston Globe*, August 14, 1987, Metro 17.

65. Reid, "South Bay Trash Incinerator Nearing Gantlet."

66. Reid, "South Bay Trash Incinerator Nearing Gantlet."

67. Reginald Jones, as quoted in Reid, "South Bay Trash Incinerator Nearing Gantlet."

68. Sarah A. Moore, "Garbage Matters: Concepts in New Geographies of Waste," *Progress in Human Geography* 36, no. 6 (December 2012): 780–799, https://doi.org/10.1177/0309132512437077.

69. EDIC Proposal for Development of Markets for Local Recyclable Materials, July 1991, BMA/0246.001/Flynn-Issues-Development-Policy-Tourism/Recycling.

70. Quill, "Boston to Truck Trash."

71. Citizen Advisory Committee on Solid Waste Disposal, "Issues to Consider."

72. Quill, "Trash Plan Meeting Opposition."

73. Letter from Bruce Hendrickson to Joseph Casazza, July 31, 1985, and letter from Bruce Hendrickson to Joseph Casazza, May 2, 1986, BMA/Casazza Files/Box 32/WTE/Folder—American REF-Fuel.

74. Thomas Andrew Savage, "Boston Harbor: The Anatomy of a Court-Run Cleanup," *Boston College Environmental Affairs Law Review* 22, no. 2 (December 1995); Eric Jay Dolin, *Political Waters: The Long, Dirty, Contentious, Incredibly Expensive but Eventually Triumphant History of Boston Harbor—A Unique Environmental Success Story* (Amherst: University of Massachusetts Press, 2004).

75. Savage, "Boston Harbor."

76. Matthew Wald, "Island Chosen for Boston Area Sewer Plant," *New York Times*, July 11, 1985, A16; Joan Vennochi and Michael K. Frisby, "Flynn Agrees to Site for Jail: State Offers Aid for South Bay," *Boston Globe*, May 14, 1986; Dolin, *Political Waters*; Savage, "Boston Harbor."

77. Alden Raine, personal interview, May 13, 2016; Sullivan, personal interview.

78. Frank Keefe, personal interview, August 30, 2014; Sullivan, personal interview; Vennochi and Frisby, "Flynn Agrees to Site for Jail."

79. Sullivan, personal interview.

80. Keefe, personal interview; Sullivan, personal interview.

81. There is of course also the ethical consideration of locating a prison adjacent to a toxic facility, but there is no evidence that this was a concern at the time.

82. Anna R. Davies, "Incineration Politics and the Geographies of Waste Governance: A Burning Issue for Ireland?," *Environment and Planning C: Government and Policy* 23 (2005): 357–397.

83. Bruce Mohl, "Bulger Stuns Hub Officials by Opposing Incinerator," *Boston Globe*, July 16, 1987, B1.

84. M. E. Malone, "A Storm on the State House Steps," *Boston Globe*, July 18, 1987, B1.

85. Bruce Mohl, "Bulger Proposes Weston Trash Site: Plan Is Alternative to Hub Incinerator," *Boston Globe*, September 17, 1987, B1; Richard Saltus, "EPA, Others Wary of Incinerator-at-Sea Plan," *Boston Globe*, September 19, 1987, B1.

86. Larry Tye, "No Easy Task to Find Site for Trash Facility," *Boston Globe*, July 22, 1987, B1; Mohl, "Bulger Proposes Weston Trash Site"; Mohl and Malone, "Bulger Looks Outside Boston"; Peter Sleeper, "Bulger Wants Advisory Panel to Seek New Incinerator Sites," *Boston Globe*, July 31, 1987, Metro 57; Frank Phillips, "Rezoning in Weston Kills Trash Proposal," *Boston Globe*, May 11, 1988, Metro 42.

87. Bruce Mohl, "Flynn Says Bulger Is Stalling Bills to Aid City," *Boston Globe*, July 21, 1987, B1; Bruce Mohl, "Bulger Amendment: Confusion, No Debate," *Boston Globe*, July 23, 1987, B1.

88. Malone, "Storm on the State House Steps."

89. M. E. Malone and Bruce Mohl, "Trash Debate Heats Up," *Boston Globe*, July 17, 1987, B1.

90. Sullivan, personal interview.

91. Daniel Golden, "A 'Turf War' Widens: South Boston Neighbors Flynn and Bulger Are Really Miles Apart," *Boston Globe*, July 26, 1987, Metro 69; Peter Howe, "Bulger Gets Mixed Reviews, Some Residents Criticize Incinerator Move," *Boston Globe*, July 19, 1987, Metro 21; "Undermining an Incinerator," editorial, *Boston Globe*, July 17, 1987, A12; "Incinerator Insight."

92. While this was all playing out, Massachusetts was in the process of creating the MWRA and planning, and then building, the new sewage treatment facility—so it is not the case that politics debilitated all infrastructure planning. The creation of the MWRA and the construction of the Deer Island plant were controversial, highly technical, politically difficult moves. But both city and state were focused on them; the stakes were high, and many actors coalesced to make it a priority. This never happened for the incinerator.

93. Peggy Hernandez, "Recycling Plan Proposed by Flynn to Dispose of City's Trash," *Boston Globe*, July 27, 1988, B17.

94. Christopher B. Daly, "Greenpeace Delivers Message to EPA: Incinerator Ash Is Toxic," Associated Press, December 12, 1987.

95. Sullivan, personal interview.

96. Letter from David Ozonoff to Lewis Pollack, commissioner of health and hospitals, May 2, 1986, BMA/0246/Box V-9/1-2.

97. Memorandum from Peter Watson to Joseph Casazza, "Quincy Residue Disposal Assessment," April 28, 1987, BMA/Casazza Files/Box 32/American REF-Fuel, Inc.

98. City of Boston, Request for Developers of a Waste-to-Energy Project, April 25, 1983, BPL/Gov Docs/M3.B16/PW/83.2.

99. Memorandum from John L. Culp to Dan Harkins, October 29, 1985, BMA/Casazza Files/Box 32/CSI.

100. Barry Commoner, "Excerpt from Meeting of Special Committee on Waste Management," September 28, 1988, BMA/0246.001/Box V-9/4-4.

101. Citizen Advisory Committee, "Issues to Consider."

102. Robin Nagle, "Garbage: Learning to Unsee" (Lecture, September 14, 2011).

103. Dominique Laporte, *History of Shit* (Cambridge, MA: MIT Press, 2000).

104. Sullivan, personal interview.

Chapter 3. Deconstructing Garbage: Radical Reframing in Seattle

1. Matthew Klingle, *Emerald City: An Environmental History of Seattle* (New Haven, CT: Yale University Press, 2007).

2. 1983 Annual Report of the Seattle Engineering Department, Seattle Municipal Archives (SMA hereafter)/1802/G2 Engineering Dept. Annual Reports, 3.

3. Klingle, *Emerald City*.

4. Over the years, the organizational structure of Seattle's solid waste department has shifted several times; sometimes it has been a subunit of the Engineering or Public Works Department, sometimes its own department overseen directly by Seattle Public Utilities (SPU). But the people who work there generally refer to it simply as "the Utility," or the Solid Waste Utility, and so that is how I refer to it in this chapter.

5. Resolution 25872, City of Seattle, 1978. It should be noted that recycling and composting were relatively new areas for municipal intervention at the time; there was no tradition of planning for these activities, nor were there comparable programs in other places to serve as models.

6. Letter from Sam Sperry, director of the Energy Department, to Paul Kraabel, president of the city council, January 15, 1980, SMA/4675-02/Solid Waste, Composting 1980 1-6.

7. "News Release: Composting Information Available," Seattle Engineering Department, 1980, SMA/4675-02 Solid Waste, Composting 1980 1-6.

8. Letter from Sam Sperry to Paul Kraabel, January 22, 1980, SMA/4601-02/Box 23/Utilities-Solid Waste-Energy Recovery I.

9. Letter from Sam Sperry to Paul Kraabel, January 22, 1980.

10. "Proposed Composting and Waste Reduction Strategy," Seattle Solid Waste Utility, 1981.

11. "Proposed Composting and Waste Reduction Strategy."

12. Candace Dempsey, "Who Gets the Garbage?," *Argus*, December 5, 1980, 1.

13. Dempsey, "Who Gets the Garbage?"

14. "Put Garbage In, Get Power out," *Seattle Times*, May 11, 1980.

15. Susan Goldberg, "City Looks at Garbage for Energy," *Seattle Post-Intelligencer*, December 22, 1982.

16. The city council formally adopted a 40 percent goal by 2010 in Resolution 27503 in 1986. This resolution was taken on the basis of the 1986 "Solid Waste Management Study: Policy and Development Plan" (SWMS) as well as several years of planning debates.

17. A 1989 EPA report estimated that the United States recycled only about 10 percent of its municipal waste in total. The report, which, incidentally, included the interlocking oval ISWM diagram explored in the introduction, also featured a dozen municipalities with ongoing recycling programs, including Seattle. Austin, Texas, recycled about 0.8 percent of its curbside waste in 1982; a small suburb of Buffalo, New York, recycled 25 percent of its waste stream in 1981; some counties with mandatory recycling in New Jersey were also reporting 25–30 percent recycling rates in the mid-1980s. Even in comparison to the most aggressive programs at the time, 40 percent was ambitious "Recycling Works: State and Local Solutions to Solid Waste Management Problems," EPA, Office of Solid Waste, January 1989.

18. Zsuzsa Gille, *From the Cult of Waste to the Trash Heap of History: The Politics of Waste in Socialist and Postsocialist Hungary* (Bloomington: Indiana University Press, 2007); Anna R. Davies, *The Geographies of Garbage Governance: Interventions, Interactions, and Outcomes* (Aldershot, UK: Ashgate, 2008); Dietmar Offenhuber, David Lee, Malima I. Wolf, and Santi Phithakkitnukoon, "Putting Matter in Place," *Journal of the American Planning Association* 78, no. 2 (2012): 173–196.

19. Memorandum from city council staff member Paul Matsuoka to all councilmembers, "Divided Report—Solid Waste Comprehensive Plan," September 5, 1986; letter from Mayor Charles Royer to Sam Smith, city council president, May 28, 1986, SMA/4601-01/4-5.

20. Letter from Sam Sperry to Paul Kraabel, January 22, 1980.

21. Letter from Mayor Royer to Sam Smith, May 28, 1986.

22. Letter from Sam Sperry to Paul Kraabel, January 22, 1980.

23. Memorandum from Paul Matsuoka to all councilmembers, September 5, 1986.

24. "Action on Solid Waste Comprehensive Plan," City Council Environmental Management Committee, September 1986, SMA/4601-01/4-5, 2; memorandum from city council staff member Nancy Glaser to all councilmembers, "Executive Summary of Environmental Management Committee's Recommendations on Recycling/Disposal Options and Solid Waste Rates," September 30, 1986, SMA/4601-01/4-4.

25. "Recommendations to the Mayor: A System to Manage Solid Waste," Solid Waste Utility, 3-4, SMA/2600-05/5-8.

26. Memorandum from Tom Tierny to Virginia Galle, Re: Response to Council Staff Information Request, August 20, 1987, SMA/2600-05/5-8; "Solid Waste Utility Recommendations to the Mayor."

27. Letter from Mayor Royer to Sam Smith, August 3, 1987, SMA/4601-01 Solid Waste Comprehensive Plan (2) 1984-1988/4-5.

28. Letter from Mayor Royer to Sam Smith, August 3, 1987.

29. Letter from Nancy Luenn to Mayor Royer, July 31, 1987, SMA/4630-02/Energy—Resource Recovery—Letters and Info 1987/7-11.

30. "Statement on the City of Seattle's Disposal Options Project," press release, Municipal League of Seattle/King County and the League of Women Voters, August 11, 1987, SMA/4630-02/Energy—Resource Recovery—Letters and Info 1987/7-13.

31. Letter from Mayor Royer to Audrey Gruger, August 11, 1987, SMA/2602-02/34-11.

32. Letter from Tim Hill to Mayor Royer, August 14, 1987, SMA/2600-05/5-8.

33. Jennifer Bagby, "The Past, Present and Future of Full Cost Accounting in Solid Waste Management," Seattle Public Utilities, 1999.

34. Serin D. Houston, *Imagining Seattle: Social Values in Governance*, Our Sustainable Future (Lincoln: University of Nebraska Press, 2019).

35. Seattle Solid Waste Pilot Programs Source Separation and Variable Rate Study, Seattle Solid Waste Utility, October 1979, Seattle Public Library (SPL hereafter)/SEADOC/E5.9.

36. Susan Gilmore, "Variable-Rate Garbage Fees Approved by Seattle Council to Start January 1," *Seattle Times*, June 17, 1980, C1; Susan Gilmore, "Garbage Rates May Go Up 70 Cents," *Seattle Times*, September 15, 1981, C2; Marshall Wilson, "Garbage Rates Going Up to Pay Cost of Fixing Midway Landfill," *Seattle Times*, August 30, 1985, B3; Susan Gilmore, "Garbage Rates Will Skyrocket in Seattle—Increases from 59 Percent to 82 Percent in Effect Aug. 1," *Seattle Times*, July 22, 1986, B1; Scott Maier, "Our Garbage Rates Could Increase 80 Percent," *Seattle Post-Intelligencer*, January 20, 1986, A3; Scott Maier, "Garbage Pickup Rates Would Skyrocket Under Proposal by the Mayor," *Seattle Post-Intelligencer*, April 23, 1986, A1; David Schaefer, "Council OK's 60 Percent Boost in Garbage Rates," *Seattle Times*, July 4, 1986, B1; David Schaefer, "Boost in Garbage Rates Called Unfair," *Seattle Times*, May 7, 1986, D1; Bob Lane and David Schaefer, "Royer Asks 79 Percent Boost in Trash Fees," *Seattle Times*, April 22, 1986, A1; Don Hannula, "Today's Seattle Rate Increase New Challenge for the Garbage Gestapo," op/ed, *Seattle Times*, August 1, 1987.

37. Letter from Eugene Avery to Mayor Royer, May 23, 1986, SMA/2602-02/1986/34-10; "Solid Waste Utility Recommendations to the Mayor."

38. "Waste to Energy: Seattle Engineering Department Energy Recovery Newsletter," February 1984, SMA/2600-06/1-2.

39. Memorandum from Eugene Avery, director of engineering, to Councilmember Virginia Galle, April 23, 1986, SMA/2602-02/34-7.

40. Susan Goldberg, "City Keeps Kent Dump, But Critics Say It Stinks," *Seattle Post-Intelligencer*, October 19, 1982, C1.

41. David Schaefer, "No Tears: Neighbors Celebrate Closing of Landfill," *Seattle Times*, October 1, 1983, A9; David Suffia, "Kent Demands That Seattle Close 2 Garbage Landfill Sites," *Seattle Times*, July 0, 1903, C1.

42. Duff Wilson, "Midway Dump Is Closed but Controversy over Landfill's Future Cleanup Continues," *Seattle Post-Intelligencer*, October 1, 1983, A3.

43. Schaefer, "No Tears."

44. Postcard from Seattle resident Nancy Jo Rauch to Virginia Galle, September 6, 1987, SMA/4630-02/7-13; Charles E. Brown, "Methane at Midway Landfill Still Keeps Two Families from Homes," *Seattle Times*, December 5, 1985, B4; Jane Hadley, "Neighbors of Landfill Blast City and State," *Seattle Post-Intelligencer*, December 5, 1985, D2; John Harris, "City Moves to Set the Terms for Buying Homes Near Landfill," *Seattle Post-Intelligencer*, March 26, 1866, A3; Kathy Bunnell Johnson, "Get

Rid of Gas Peril at Landfill, Seattle Warned," *Seattle Post-Intelligencer*, June 29, 1985, A3; Maier, "Our Garbage Rates Could Increase"; Scott Maier, "Seattle to Purchase Landfill for $1 Million," *Seattle Post-Intelligencer*, April 5, 1986, A1; Elizabeth Pulliam, "Methane Jeopardizes Day-Care Center," *Seattle Times*, August 29, 1985, D1; Richard Seven, "Easing Fears Near Midway Landfill," *Seattle Times*, April 2, 1986, D1; Wilson, "Garbage Rates Going Up."

45. Letter from Mayor Royer to Sam Smith, May 26, 1986, SMA/4601-01/Solid Waste Comprehensive Plan (2) 1984–1988/4–5.

46. For example, as one resident put it, "The proposed Southend [*sic*] site is stepping on my toes, I live here. We of the Southend [*sic*] have had many crosses to bear in the past ten year [*sic*]. Robberies, Rapes, Assaults, Murders, Rock Houses and now THIS. Really we do not need any more." Letter from June E. Cappetto to Councilman George Benson, September 6, 1987, SMA/4630-02/7-13.

47. Letter from Richard Vincent to the Seattle City Council, July 29, 1987, SMA/4630-02/Energy—Resource Recovery—Letters and Info 1987/7-11.

48. Letter from Mr. and Mrs. Frank Barker to Councilwoman Virginia Galle, July 28, 1987, SMA/4630-02/Energy—Resource Recovery—Letters and Info 1987/7-11; Petition from South End Seattle Community Organizations (SESCO) with 450 signatures, to Mayor Charles Royer, July 31, 1987, SMA/4630-02/Energy—Resource Recovery—Letters and Info 1987/7-12.

49. Laura Pulido, Steve Sidawi, and Robert O. Vos, "An Archaeology of Environmental Racism in Los Angeles," *Urban Geography* 17, no. 5 (July 1996): 419–439, https://doi.org/10.2747/0272-3638.17.5.419; Robert D. Bullard, *Dumping in Dixie: Race, Class and Environmental Quality* (Boulder: Westview Press, 1994).

50. Letter from George Park, South Park Community Club, to Councilwoman Virginia Galle, August 13, 1987, SMA/4630-02/Energy—Resource Recovery—Letters and Info 1987/7-12.

51. Letter to Mayor Royer from Curtis Nimz, July 27, 1987, SMA/4630-02/Energy—Resource Recovery—Letters and Info 1987/7-11.

52. "Don't Let Seattle Get Burned," flyer, sent to the city council from Penny Danna, 1987, SMA/4630-02/7-13.

53. "Use it Again Seattle," press release, May 5, 1980, SMA/4675-02 Solid Waste, Composting 1980/1-6.

54. Robert A. Lowe and James P. McMahon, "Proposed Recycling and Waste Reduction Strategy for the City of Seattle," vol. 1, Seattle Engineering Department, January 7, 1981, SPL/353.93216 L9516P 1981.

55. Mike Sato, "Garbage: Getting a Grip on Solid Waste," *Seattle Sun*, March 26, 1980, 8.

56. "Solid Waste Utility Recommendations to the Mayor."

57. "Solid Waste Management Study: Policy and Development Plan," Seattle Engineering Department, January 1986, SMA/Report #3127.

58. Recycling was broadly supported in the city at this time. The Utility, the city council, and the mayor were receiving regular constituent letters urging the city to consider aggressive recycling as an alternative to incineration. The League of Women Voters and the Seattle Municipal League captured the sentiment of many in their statement to Mayor Royer. While noting that an incinerator in Seattle would likely be necessary, they urged the city to "intensify its efforts in terms of time, money and

planning resources in the areas of waste reduction (such as packaging legislation), recycling, and composting before making the critical decision on the potential size of any incinerator." They continued: "This implies the need for solid waste management systems which maximize the ease and efficiency of waste reduction and recycling. The greater the extent to which materials are not introduced into the waste stream, or once introduced are recycled, the smaller the size and cost of any incinerator. We find ourself [*sic*] in agreement with Mayor Royer that the goal of a 1 percent annual reduction in the waste stream over the next twenty years is overly modest and lacks the aggressiveness which the city has consistently shown in its planning efforts for an incinerator." "Statement on the City of Seattle's Disposal Options Project."

59. Letter from Ed Hebbert and Martha Taylor to Virginia Galle, July 27, 1987, SMA/4630-02/7-11.

60. Letter from Nancy Luenn to Mayor Royer, July 31, 1987.

61. In fact, WTE is still—under US policy—considered a form of "renewable" energy. This characterization is of course suspect, because most plastics are made from natural gas and other petroleum products, and obviously are not renewable.

62. Letter from Kathy Jordan to Mayor Royer, August 5, 1987, SMA/4630-02/Energy—Resource Recovery—Letters and Info 1987/7-14.

63. Letter from Sid and Laura Maroney to Mayor Royer, July 20, 1987, SMA/4630-02/Energy—Resource Recovery—Letters and Info 1987 7-11.

64. "On the Road to Recovery: Seattle's Integrated Solid Waste Management Plan," City of Seattle, August 1989.

65. "Waste Watch: A Balanced Approach," July 1984, SMA/2600-06/1-2.

66. Brown, "Methane at Midway"; Pulliam, "Methane Jeopardizes Day-Care Center"; Duff Wilson, "City Advised to Hire Gas Expert 11 Months Ago," *Seattle Post-Intelligencer*, November 4, 1986, D2; David Schaefer, "Midway: A Trail of Trouble," *Seattle Times*, February 4, 1987, A1.

67. Gilmore, "Garbage Rates May Go Up 70 Cents"; Wilson, "Garbage Rates Going Up"; Gilmore, "Garbage Rates Will Skyrocket in Seattle"; Maier, "Our Garbage Rates Could Increase"; Schaefer, "Council OK's 60 Percent Boost"; Schaefer, "Boost in Garbage Rates Called Unfair."

68. Susan Gilmore, "Audit: Pay-by-Can Rates Failing to Cut Back Garbage," *Seattle Times*, December 17, 1981, C10.

69. "City's Garbage Utility Doesn't Rate Very Well," editorial, *Seattle Times*, May 1, 1981; Eric Naider, "Ex-Manager Calls Garbage Rates a Rip-Off," *Seattle Post-Intelligencer*, May 2, 1981, A7.

70. Susan Gilmore, "Head of Garbage Utility Leaves Job," *Seattle Times*, April 20, 1981, C2.

71. Diana Gale, personal interview, January 5, 2015.

72. "Waste to Energy Newsletter," Seattle Engineering Department, September–October 1983, SMA/2600-06/Waste to Energy and Waste Watch 1983-1990/1-2.

73. Letter from Mayor Royer to Norm Rice, city council president, August 10, 1984, SMA/4601-01/Solid Waste Comprehensive Plan (2) 1984-88/4-5.

74. Memorandum from Gary Zarker, director of engineering, to Mayor Royer, June 1, 1987, 2602-02 Solid Waste Pilot Program 1986-87 33-14.

75. "Solid Waste Management Study: Policy and Development Plan," vol. 2, "Final

Report to the City Council," Seattle Solid Waste Utility, May 1986, SMA/4601-01 Solid Waste Comprehensive Plan (2) 1984-1988/4-5.

76. Letter from Al Yamagiwa via Jerry Garman to Randy Hardy, Seattle Engineering Department, April 17, 1986; letter from Randy Hardy to Eugene Avery, director of engineering, April 21, 1986, 2602-02 Solid Waste Recycling 1986 34-7.

77. Alice Lui, Paul Matsuoka, and Bob Morgan, "City Council Staff Analysis of Solid Waste Management Study," vol. 2, "Policy and Development Plan," August 12, 1986, SMA/2600-05/5-10a.

78. Environmental Management Committee's Action on Solid Waste Comprehensive Plan, September 5, 1986, SMA/4601-01/Solid Waste Comprehensive Plan (2) 1984-1988/4-5; letter from Diana Gale to Councilmembers Jim Street and Jane Noland, September 27, 1987, SMA/2600-05/8.

79. It must be noted that this client–service provider relationship is not universal. In many small cities and towns, people still bring their garbage to a dump or transfer station. Some of these places in Massachusetts, such as Acton and Wellesley, have developed sophisticated recycling and reuse programs where residents separate their wastes into highly specific categories—which is much more effective than single-stream collection programs. In large cities where most people live in multifamily buildings, people may drop their waste into common dumpsters, and then building staff are responsible for its removal. The client-service provider relationship, though, was dominant in Seattle and many midsize cities where the bulk of residents lived in single-family or small multifamily housing and so had a direct relationship with garbage collection.

80. Letter from Mayor Royer to Sam Smith, October 11, 1988, SMA/4601-01/4-4; memorandum from Diana Gale to Councilmember George Benson, September 15, 1988, SMA/4601-01/4-4.

81. Joan Edwards, quoted in Pete McConnell, "Peer Pressure Holds Key to Successful Recycling Attempts," *Seattle Post-Intelligencer*, September 25, 1986.

82. "Recycling Works"; "Sustainable Materials Management: The Road Ahead," EPA, June 2009, www.epa.gov/smm/sustainable-materials-management-road-ahead.

83. Letter from Mayor Royer to Sam Smith, May 28, 1986.

84. Letter from Diana Gale to George Benson, September 15, 1988.

85. Letter from Mayor Royer to Sam Smith, September 22, 1988, SMA/4601-01 Solid Waste Comprehensive Plan (1) 1984-1988/4-4.

86. Resolution 27867, City of Seattle, 1987.

87. Tim Croll, personal interview, January 8, 2015.

88. Letter from Mayor Royer to Sam Smith, October 11, 1988.

Chapter 4. Compliant and Defiant Wasteways: Boston and Seattle within the WRWR

1. Gregory C. Unruh, "Understanding Carbon Lock-In," *Energy Policy* 28, no. 12 (October 2000): 817–830, https://doi.org/10.1016/S0301-4215(00)00070-7; Frans Berkhout, "Technological Regimes, Path Dependency and the Environment," *Global Environmental Change* 12 (2002): 1–4; Lea Fuenfschilling and Bernhard Truffer, "The Structuration of Socio-Technical Regimes—Conceptual Foundations from Institu-

tional Theory," *Research Policy* 43, no. 4 (May 2014): 772–791, https://doi.org/10.1016/j
.respol.2013.10.010.

2. Raymond L. Flynn, "Statement of Raymond L. Flynn on Recycling Ordinance,"
July 25, 1990, BMA/Collection 0246.001/Box V-9/Environment: Recycling 2.

3. Peter Howe, "Flynn, Scondras Disagree over Need for Fees, Fines in Recycling
Plan," *Boston Globe*, December 12, 1988, Metro 19.

4. Neil Sullivan, personal interview, May 9, 2016.

5. Steve Marantz, "Mayor, Environmentalists Spar on Plan for Curbside Recycling," *Boston Globe*, July 25, 1990.

6. Flynn, "Statement of Raymond L. Flynn on Recycling Ordinance"; "Recycling
Program," § 7-13, City of Boston, 1990.

7. Steve Marantz, "City OK's Curbside Recycling Ordinance," *Boston Globe*, July
26, 1990.

8. Marantz, "City OK's Curbside Recycling"; "Recycling Program," § 7-13.

9. Letter from Donald Gillis to Mike Lindberg, July 11, 1991, BMA/Flynn-Issues-
Development-Policy-Tourism/Recycling; memorandum from Rick Innes to "Distribution," December 4, 1991, Proposal Evaluation, RFP for acceptance of used newspaper, July 1, 1991–June 30, 1994, undated, BMA/Casazza Files/Box 56/Recycling
1991.

10. Letter from Rick Innes, PWD, to the board members of the Boston Food Coop,
September 18, 1991, BMA/Casazza Files/Box 56/Recycling 1991.

11. Letter from Joseph P. Casazza to Howard Waddell, director of marketing for
the Massachusetts Bay Transportation Authority (MBTA), November 16, 1991; Boston Recycling Advisory Committee Meeting Minutes, September 18, 1991, BMA/
Casazza Files/Box 56/Recycling 1991.

12. Letter from Margaret Toro to Mayor Flynn, October 25, 1991; letter from
Joseph Casazza to Margaret Toro, November 12, 1991, BMA/Casazza Files/Box 56/
Recycling 1991.

13. Letter from Joseph Casazza to Mayor Flynn, October 22, 1990, BMA/Casazza
Files/Box 56/Recycling 1991.

14. Helen Kennedy, "Hub Gets Trashed in Recycling Survey," *Boston Herald* (date
not included in news clipping), BMA/Casazza Files/Box 56/Recycling 1991.

15. Letter from Marya P. Labarthe to Joseph Casazza, January 3, 1991; letter from
the Jamaica Plain Recycling Task Force to Joseph Casazza, February 1, 1991; letter
from Cynthia Grubbs to Joseph Casazza, February 7, 1991; letter from Julie McVay to
Stan Gross, April 8, 1991, BMA/Casazza Files/Box 56/Recycling 1991.

16. Jenna Nierstedt, "Boston to Begin No-Sort Recycling Program July 1," *Boston
Globe*, June 24, 2009, B13.

17. The following discussion shares some primary source material with Lily Baum
Pollans, "Trapped in Trash: 'Modes of Governing' and Barriers to Transitioning to
Sustainable Waste Management," *Environment and Planning A*, July 12, 2017, https://
doi.org/10.1177/0308518X17719461.

18. Clarissa Morakawski, "Single Stream Uncovered," *Resource Recycling* 3 (2010):
17–39; Eureka Recycling, "The Downstream of Single Stream," *Resource Recycling*,
November 2002, www.eurekarecycling.org/pdfs/ResourceRecyclingArticle.pdf; Dan
Emerson, "Single Stream vs. Source Separated Recycling," *BioCycle* 45, no. 3 (March

2004): 22–25; "Download: Single Stream Carts Timeline," Recycling Partnership, March 9, 2017, https://recyclingpartnership.org/single-stream-carts-timeline; "Let's Roll: A Webinar on the How's and Whys of Single Stream Cart Recycling," Recycling Partnership, May 19, 2015, https://recyclingpartnership.org/lets-roll-a-webinar-on -the-hows-and-whys-of-single-stream-cart-recycling.

19. Boston Public Works Employee, personal interview, January 31, 2012.

20. Matt Carroll, "Recycling Becomes Cash Cow: New Systems Gain Popularity, Profits," *Boston Globe*, September 11, 2008, GS1; Boston Public Works Employee, personal interview.

21. "Mayor's Performance Report: Boston Public Works and Transportation Departments," Boston About Results, City of Boston, December 2009, www.cityofboston .gov/Images_Documents/PWD-BTD_BAR percent20FY10 percent20Q2_tcm3-177 83.pdf.

22. Unruh, "Understanding Carbon Lock-In."

23. Thomas P. Hughes, *Networks of Power: Electrification in Western Society, 1880– 1930* (Baltimore: Johns Hopkins University Press, 1983).

24. Resolution 30990, City of Seattle, 2007.

25. Resolution 27871, City of Seattle, 1988; Resolution 29805, City of Seattle, 1998; Resolution 30750, City of Seattle, 2005; Resolution 31426, City of Seattle, 2013.

26. "2015 Recycling Rate Report," Seattle Public Utilities, July 1, 2016, www .seattle.gov/util/cs/groups/public/@spu/@garbage/documents/webcontent/1_052510 .pdf.

27. "2018 Waste Reduction and Recycling Report," Seattle Public Utilities, July 1, 2019, www.seattle.gov/Documents/Departments/SPU/Documents/Recycling_Rate_ Report_2018.pdf.

28. "Seattle's Solid Waste Plan: On the Path to Sustainability," Seattle Public Utilities, August 1998, SMA/Report #4285.

29. "Picking Up the Pace to Zero Waste: Seattle's Solid Waste Plan, 2011 Revision," Seattle Public Utilities, 2013.

30. Resolution 27867, City of Seattle, 1987.

31. Resolution 27871.

32. Tim Croll, personal interview, January 8, 2015.

33. Aaron Spencer, "Ins and Outs of Recycling," *Seattle Times*, March 1, 2015, B1; "2015 Recycling Rate Report."

34. Art James, "Letter to the Editor: Unsettling Difference—Why Doesn't Tacoma Urge Garbage Recycling?," *Seattle Times*, January 23, 1995, B3; Ray Hoffman, personal interview, January 13, 2015.

35. Aaron Spencer, "A Case Study in Trash Sorting: Inspecting the Bins at the Seattle Times," *Seattle Times*, August 13, 2015, www.seattletimes.com/seattle-news /going-through-the-trash-in-the-seattle-times-breakroom.

36. Don Hannula, "A Stacked Deck? Tiptoeing Toward the Garbage Curb," *Seattle Times*, November 7, 1986, A10; Don Hannula, "Today's Seattle Rate Increase New Challenge for the Garbage Gestapo," op/ed, *Seattle Times*, August 1, 1987; Don Hannula, "No Wonder We Do the Garbage Can Stomp," op/ed, *Seattle Times*, February 6, 1987; Don Hannula, "Warning: The Garbage Gauge Cometh," op/ed, *Seattle Times*, June 12, 1991; Don Hannula, "The Garbageman Cometh (Whenever He Pleases),"

op/ed, *Seattle Times*, September 22, 1993, http://community.seattletimes.nwsource.com /archive/?date=19931028&slug=1728387; Don Hannula, "What's Next for Seattle: A Ph.D. in Composting?," *Seattle Times*, August 17, 1995, B6; Don Hannula, "Keep a Stiff Upper Lid on the Ol' Garbage Can," op/ed, *Seattle Times*, November 30, 1995.

37. Resolution 27828, City of Seattle, 1988.

38. Resolution 27980, City of Seattle, 1989.

39. Ordinance 114035, City of Seattle, 1988.

40. "City Employees Recycle," Seattle Engineering Department's Office for Citizen Participation, n.d., SMA/2600-06/1-1.

41. Resolution 28637, City of Seattle, 1992.

42. Ordinance 121372, City of Seattle, 2003.

43. Bob Young, "Recyclables Now Must Be Kept out of Garbage," *Seattle Times*, January 1, 2005, B1; Elizabeth M. Gillespie, "Recycling Effort Ratchets Up," *Seattle Times*, February 6, 2005, B6; Ordinance 121372.

44. Young, "Recyclables Now Must Be Kept out of Garbage"; Gillespie, "Recycling Effort Ratchets Up"; Ordinance 121372.

45. Susan Gilmore, "Gearing Up to Help the Planet," *Seattle Times*, June 16, 2005, B1; Jerry Large, "Waste Can Be a Terrible Thing to Mind," *Seattle Times*, July 3, 2005, L1.

46. "Yard Waste Report," Seattle Public Utilities, December 2005, www.seattle .gov/util/cs/groups/public/@spu/@garbage/documents/webcontent/COS_004534 .pdf.

47. Ordinance 124582, City of Seattle, 2014; Spencer, "Ins and Outs of Recycling."

48. This is only the case for small buildings where residents manage their own trash collection; residents in large, multifamily buildings (over five units) have dumpster arrangements that don't provide the same signals.

49. For example, David H. Folz, "Recycling Program Design, Management and Participation: A National Survey of Municipal Experience," *Public Administration Review* 51, no. 3 (1991): 222–231; Margaret A. Reams and Brook H. Ray, "The Effects of Three Prompting Methods on Recycling Participation Rates: A Field Study," *Journal of Environmental Systems* 22, no. 4 (1993): 371–379; Darren Perrin and John Barton, "Issues Associated with Transforming Household Attitudes and Opinions into Materials Recovery: A Review of Two Kerbside Recycling Schemes," *Resources, Conservation, and Recycling* 33, no. 1 (2001): 61–74.

50. "Garbage Report: 1st Quarter 2019," Seattle Public Utilities, Economic Services Division, April 30, 2019.

51. "Draft City of Seattle Solid Waste Facilities Master Plan," Seattle Public Utilities, November 2003, www.seattle.gov/UTIL/MyServices/Garbage/AboutGarbage /SolidWastePlans/FacilitiesPlan/index.htm.

52. "Draft City of Seattle Solid Waste Facilities Master Plan," "Appendix B: Public Involvement Report and Meeting Summaries," Seattle Public Utilities, November 2003.

53. Resolution 30431, City of Seattle, 2001.

54. Although Seattle is known for progressive and inclusive politics, it must be recognized that its history is imperfect in this regard. The city has consistently excluded nonwhite and low-income populations from decision-making. For a thorough analysis,

see Serin D. Houston, *Imagining Seattle: Social Values in Governance*, Our Sustainable Future (Lincoln: University of Nebraska Press, 2019).

55. Carl Zimring, "Dirty Work: How Hygiene and Xenophobia Marginalized the American Waste Trades, 1870–1930," *Environmental History* 9, no. 1 (January 2004): 80–101, https://doi.org/10.2307/3985946; Robin Nagle, *Picking Up: On the Streets and Behind the Trucks with the Sanitation Workers of New York City* (New York: Farrar, Straus and Giroux, 2013).

56. "North Transfer Station: Seattle's Newest Facility," Seattle Public Utilities, 2016, www.seattle.gov/Documents/Departments/SPU/Services/Garbage/NTSbrochure .pdf; Daniel Beekman, "Nicest Dump Around? New $108M Transfer Station in Wallingford Even Has a Basketball Court," *Seattle Times*, December 5, 2016, www .seattletimes.com/seattle-news/politics/108-million-dump-opens-in-wallingford. For more on normalizing contact with discards, see Mira Engler, *Redesigning America's Waste Landscapes* (Baltimore: Johns Hopkins University Press, 2004).

57. Mary Jean Spadafora, "Opt-Out Program Results in Fewer Phone Books," *Seattle Times*, April 24, 2014, B4.

58. Croll, personal interview.

59. Emily Heffter, "City May Hang Up on Phonebook Blitz—Ordinance Aims to Curtail Distribution," *Seattle Times*, June 23, 2010, B1; Spadafora, "Opt-Out Program"; "Stop Phone Books," Seattle Public Utilities, www.seattle.gov/util/Environ mentConservation/OurCity/ReduceReuse/StopPhoneBooks/index.htm.

60. Croll, personal interview.

61. Croll, personal interview.

62. John Kingdon, *Agendas, Alternatives, and Public Policies*, 2nd ed. (New York: Harper Collins College, 1995).

63. Jon Savelle, "Debate over Styrofoam Won't Fade Away," *Seattle Times*, February 12, 1991, B7; Sharon Pian Chan, "Fee and Ban? It's Time to Talk," *Seattle Times*, July 8, 2008; Noelene Clark, "Council Panel OKs Bag Fees—Foam-Container Ban Also Approved," *Seattle Times*, July 23, 2008; "Food Package Requirements," Seattle Public Utilities, www.seattle.gov/util/forbusinesses/solidwaste/foodyardbusinesses/co mmercial/foodpackagingrequirements.

64. Keith Ervin, "Seattle Considers Switch to Biweekly Garbage Pickup," *Seattle Times*, August 13, 2010, B1; Lynn Thompson, "In Seattle, Garbage Pick-Up Just Twice a Month?," *Seattle Times*, April 2, 2012, A1.

65. Lynn Thompson, "Mayor Curbs Plan to Collect Garbage Every Other Week," *Seattle Times*, February 22, 2014, B1.

66. Thompson, "Mayor Curbs Plan."

67. "Less Frequent Garbage Pickup Needs Sorting Out," editorial, *Seattle Times*, December 27, 2013, A17.

68. Cassandra Profita, "Seattle Considers Fining Residents for Failing to Compost," Oregon Public Broadcasting, February 18, 2015, www.opb.org/news/article /seattle-considers-fines-to-enforce-curbside-compos; Vanessa Ho, "New Seattle Law: No Food in Trash," *Seattle Post-Intelligencer*, December 31, 2014, www.seattlepi.com /local/article/New-Seattle-law-No-more-food-in-trash-5983805.php.

69. "About the Pacific Legal Foundation," Pacific Legal Foundation, 2016, www .pacificlegal.org/about.

70. Erik Lacitis, "Judge: Seattle Trash-Check Ordinance 'Unconstitutional,'" *Seattle Times*, April 27, 2016, www.seattletimes.com/seattle-news/politics/judge-seat tle-trash-check-ordinance-unconstitutional.

71. Sara Bernard, "Seattle's Food Waste Law Is Working, and It's Not Because of the Garbage Snooping," *Seattle Weekly*, April 29, 2016.

72. Kirk Johnson, "Residents Sue Seattle, Saying New Trash Rules Violate Privacy," *New York Times*, July 17, 2015, www.nytimes.com/2015/07/18/us/residents-sue -seattle-saying-new-trash-rules-violate-privacy.html; Dan Springer, "Seattle Sued over Recycling Inspectors Keeping Tabs on Residents' Trash," Fox News, September 14, 2015, www.foxnews.com/politics/2015/09/14/trash-talk-seattle-war-on-waste.html.

73. Quoted in Bernard, "Seattle's Food Waste Law Is Working."

74. Mary Douglas, *Purity and Danger: An Analysis of the Concepts of Pollution and Taboo*, new ed. (New York: Routledge, 1984), 5.

Chapter 5. Resisting Garbage

1. "2018 Waste Prevention and Recycling Report," Seattle Public Utilities, July 1, 2019, www.seattle.gov/Documents/Departments/SPU/Documents/Recycling_Rate _Report_2018.pdf.

2. The divide between single-family and multifamily residences does not map neatly onto demographics in Seattle, as many of the city's high-rise buildings are also among its most expensive real estate. Also, as Tim Croll noted in an interview, supported by recent city surveys, environmental concern in Seattle is widespread and not correlated with race or income. Residents of color and low-income residents report the same levels of support for city environmental programs as middle-class white residents.

3. "2018 Waste Prevention and Recycling Report."

4. "2018 Waste Prevention and Recycling Report"; Brian J. Love and Julie Rieland, "Covid-19 Is Laying Waste to Many US Recycling Programs," *The Conversation*, June 23, 2020, https://theconversation.com/covid-19-is-laying-waste-to-many-us-re cycling-programs-139733.

5. Max Liboiron, "Solutions to Waste and the Problem of Scalar Mismatches," *Discard Studies* (blog), February 10, 2014, https://discardstudies.com/2014/02/10 /solutions-to-waste-and-the-problem-of-scalar-mismatches; Max Liboiron, "Against Awareness, for Scale: Garbage Is Infrastructure, Not Behavior," *Discard Studies* (blog), January 23, 2014, https://discardstudies.com/2014/01/23/against-awareness-for-scale -garbage-is-infrastructure-not-behavior; Samantha MacBride, *Recycling Reconsidered: The Present Failure and Future Promise of Environmental Action in the United States* (Cambridge, MA: MIT Press, 2012); Timothy W. Luke, *Ecocritique: Contesting the Politics of Nature, Economy, and Culture* (Minneapolis: University of Minnesota Press, 1997); Kenneth A. Gould, Allan Schnaiberg, and Adam S. Weinberg, *Local Environmental Struggles: Citizen Activism in the Treadmill of Production* (Cambridge, UK: Cambridge University Press, 1996).

6. "Picking Up the Pace to Zero Waste: Seattle's Solid Waste Plan, 2011 Revision," Seattle Public Utilities, 2013, 1.3.

7. Tim Croll, personal interview, January 8, 2015.

8. "2018 Waste Prevention and Recycling Report."

9. "National Overview: Fact and Figures on Materials, Wastes and Recycling," EPA, March 13, 2020, www.epa.gov/facts-and-figures-about-materials-waste-and-recycling/national-overview-facts-and-figures-materials.

10. "Solid Waste Storage and Access for New or Remodeled Buildings," Seattle Public Utilities, n.d., www.seattle.gov/utilities/construction-resources/collection-and-disposal/storage-and-access.

11. "Picking Up the Pace to Zero Waste."

12. Sinnott Murphy and Stephanie Pincetl, "Zero Waste in Los Angeles: Is the Emperor Wearing Any Clothes?," *Resources, Conservation and Recycling* 81 (December 2013): 40–51, https://doi.org/10.1016/j.resconrec.2013.09.012.

13. "One Bin for All," City of Houston, 2019, www.houstontx.gov/onebinforall; "Annual Report: New York City Refuse and Recycling Statistics for Fiscal Year 2019," New York City Department of Sanitation, 2019.

14. New York City's composting pilot was halted as a result of the city's budget crisis during the coronavirus pandemic of 2020; as of this writing it was scheduled to return in June 2022. New York City Department of Sanitation, "Curbside Composting Overview," n.d., https://www1.nyc.gov/assets/dsny/site/services/food-scraps-and-yard-waste-page/overview-residents-organics.

15. "One Bin for All: A Leapfrog Approach to Increase Universal Recycling Rates," Bloomberg Philanthropies, 2013, https://mayorschallenge.bloomberg.org/ideas/one-bin-for-all.

16. "One Bin for All: Progress Report," City of Houston, December 31, 2015, www.houstontx.gov/onebinforall/OBFA_Progress_Report-20151231.pdf.

17. The One Bin project was ultimately scrapped by Houston's mayor, and the city carried on with more traditional garbage, recycling, and yard-waste composting programs. Meagan Flynn, "The Long Rise and Fast Fall of the Ambitious One-Bin Recycling Program," *Houston Press*, July 13, 2017, www.houstonpress.com/news/what-happened-to-ecohub-and-houstons-one-bin-for-all-recycling-plan-9601564.

18. Harriet Bulkeley and Michele Betsill, *Cities and Climate Change: Urban Sustainability and Global Environmental Governance* (New York: Routledge, 2005).

19. Laura Parker, "We Depend on Plastic. Now, We're Drowning in It," *National Geographic*, May 16, 2018, www.nationalgeographic.com/magazine/2018/06/plastic-planet-waste-pollution-trash-crisis.

20. Again, a small sample of relevant popular media: Gary Robbins, "UCSD Discovers Surge in Plastics Pollution off Santa Barbara," *Los Angeles Times*, September 5, 2019, www.latimes.com/california/story/2019-09-04/uc-san-diego-discovers-explosion-in-plastics-products-in-seafloor-off-santa-barbara; Laura Parker, "Here's How Much Plastic Trash Is Littering the Earth," *National Geographic*, December 20, 2018, www.nationalgeographic.com/news/2017/07/plastic-produced-recycling-waste-ocean-trash-debris-environment; Katie Day and Trent Hodges, "The Link Between Fossil Fuels, Single-Use Plastics, and Climate Change," EcoWatch, May 3, 2018, www.ecowatch.com/fossil-fuels-single-use-plastics-2565595371.html; Tom Watson, "Bathroom Shelf May Be a Source of Microplastics in Sound," *Seattle Times*, October 22, 2011.

21. Deirdre McKay, "Fossil Fuel Industry Sees the Future in Hard-to-Recycle Plastic," *The Conversation* (blog), October 10, 2019, http://theconversation.com/fossil-fuel-industry-sees-the-future-in-hard-to-recycle-plastic-123631; Michael Cork-

ery, "A Giant Factory Rises to Make a Product Filling Up the World: Plastic," *New York Times*, August 12, 2019, www.nytimes.com/2019/08/12/business/energy -environment/plastics-shell-pennsylvania-plant.html; Zöe Schlanger, "A New Wave of Plastic Is Coming to Our Shores," *Audubon*, July 29, 2020, www.audubon.org /magazine/summer-2020/a-new-plastic-wave-coming-our-shores.

22. "Plastics: Material-Specific Data," EPA, September 12, 2017, www.epa.gov /facts-and-figures-about-materials-waste-and-recycling/plastics-material-spe cific-data.

23. "#BeatPlasticPollution This World Environment Day," United Nations Environment Programme, accessed December 17, 2019, www.unenvironment.org/inter active/beat-plastic-pollution.

24. "Plastics: Material-Specific Data."

25. Parker, "Here's How Much Plastic Trash."

26. "Plastics in the Ocean," Ocean Conservancy, March 7, 2017, https://ocean conservancy.org/trash-free-seas/plastics-in-the-ocean; Cristina Files, "The Impact of the Plastic Bag Ban," CalRecycle, *In the Loop* (blog), January 9, 2017, www.calrecycle.ca .gov/blogs/in-the-loop/in-the-loop/2017/01/09/the-impact-of-the-plastic-bag-ban.

27. Ginger Hervey, "The Plastic in Our Bodies," *Politico* (blog), May 5, 2019, www .politico.eu/article/the-plastic-in-our-bodies-health; Damian Carrington, "People Eat at Least 50,000 Plastic Particles a Year, Study Finds," *Guardian*, June 5, 2019, www.theguardian.com/environment/2019/jun/05/people-eat-at-least-50000-plastic -particles-a-year-study-finds.

28. Laura Parker, "The World Agrees There's a Plastic Waste Crisis—Can It Agree on a Solution?," *National Geographic*, March 25, 2019, www.nationalgeographic.com /environment/2019/03/un-environment-plastic-pollution-negotiations.

29. "Bag the Ban: Say No to Bans and Taxes on Your Grocery Bags," American Progressive Bag Alliance, Bag the Ban, 2019, www.bagtheban.com; Kate Anderson, "To Ban or Not to Ban: The Politics of Plastic Bag Laws," Population Education, *PopEd Blog* (blog), March 30, 2018, https://populationeducation.org/to-ban-or-not-to-ban -the-politics-of-plastic-bag-laws.

30. Joseph Winters, "It's Official: Reusables Are Safe During COVID-19," *Grist* (blog), June 26, 2020, https://grist.org/climate/its-official-reusables-are-safe-during -covid-19; Tony Radoszewski testimony, "Plastic Production, Pollution and Waste in the Time of Covid-19: The Life-Threatening Impact of Single Use Plastic on Human Health," House Committee on Oversight and Reform, Subcommittee on the Environment, 116th Cong., 2nd sess., July 7, 2020; Samantha Maldonado and Marie J. French, "Plastics Industry Goes After Bag Bans During Pandemic," *Politico* (blog), March 24, 2020, https://polid.co/2UP2dvs; Hiroko Tabuchi, "In Coronavirus, Industry Sees Chance to Undo Plastic Bag Bans," *New York Times*, March 26, 2020, www .nytimes.com/2020/03/26/climate/plastic-bag-ban-virus.html.

31. "Plastic Bags Are 100% Recyclable," American Progressive Bag Alliance, Bag the Ban, www.bagtheban.com/learn-the-facts/recycling.

32. "7 Things You Didn't Know About Plastic (and Recycling)," *National Geographic Society Newsroom* (blog), April 4, 2018, https://blog.nationalgeographic.org /2018/04/04/7-things-you-didnt-know-about-plastic-and-recycling.

33. Corkery, "Giant Factory Rises."

34. Michael Corkery, "Beverage Companies Embrace Recycling, Until It Costs

Them," *New York Times*, July 5, 2019, www.nytimes.com/2019/07/04/business/plastic
-recycling-bottle-bills.html.

35. "Funding Partners," Recycling Partnership, 2020, https://recyclingpartnership
.org/funding-partners; "Our Partners," Keep America Beautiful, 2020, https://kab
.org/about/partners.

36. In the United States, the federal government has always been hesitant to regulate plastics, even prior to the dramatic environmental deregulation of the Trump administration. The EPA has even refused to regulate key chemical constituents of many
plastic products, including Bisphenol A, that have been widely acknowledged to have
endocrine disrupting properties and other health impacts. "Risk Management for Bisphenol A (BPA)," Overviews and Factsheets, EPA, September 21, 2015, www.epa
.gov/assessing-and-managing-chemicals-under-tsca/risk-management-bisphenol
-bpa; Sarah A. Vogel, "The Politics of Plastics: The Making and Unmaking of Bisphe-
nol-A 'Safety,'" *American Journal of Public Health*, special issue, *Framing Health Matters*
99, no. S3 (2009): S559–566. On the impact of plastic bag regulations, see Tatiana A.
Homonoff, "Can Small Incentives Have Large Effects? The Impact of Taxes Versus
Bonuses on Disposable Bag Use," *American Economic Journal: Economic Policy* 10, no. 4
(November 2018): 177–210, https://doi.org/10.1257/pol.20150261; Tatiana Homonoff, Lee-Sien Kao, and Christina Seybolt, "Skipping the Bag: Assessing the Impact
of Chicago's Tax on Disposable Bags," September 2018, www.ideas42.org/wp-content
/uploads/2018/09/Bag_Tax_Paper_final.pdf; "Plastic Bag Bans and Fees," Surfrider
Foundation, accessed October 5, 2016, www.surfrider.org/pages/plastic-bag-bans
-fees; "Anacostia Watershed Trash Reduction Plan," Anacostia Watershed Association, December 2008, https://doee.dc.gov/sites/default/files/dc/sites/ddoe/publica
tion/attachments/2009.01.29_Trash_Report_1.pdf; Files, "The Impact of the Plastic
Bag Ban."

37. Ryan Carra and Dacie Meng, "Federal Packaging Extended Producer Responsibility Legislation Under Development," *Beveridge and Diamond PC* (blog), August 6, 2019, www.bdlaw.com/publications/federal-packaging-extended-producer-re
sponsibility-legislation-under-development; Jennifer Nash and Christopher Bosso,
"Extended Producer Responsibility in the United States: Full Speed Ahead?," Harvard Kennedy School Mossavar-Rahmani Center for Business and Government,
2013, www.hks.harvard.edu/sites/default/files/centers/mrcbg/files/Nash_Bosso_2013
-10.pdf; "California Becomes 20th State in 2019 to Consider Right to Repair Bill,"
U.S. PIRG, March 18, 2019, https://uspirg.org/news/usp/california-becomes-20th
-state-2019-consider-right-repair-bill; "Legislation," Repair Association, July 9, 2019,
http://repair.org/legislation; "Find a Group," Buy Nothing Project, accessed December 19, 2019, https://buynothingproject.org/find-a-group.

38. On "busyness," see MacBride, *Recycling Reconsidered*. On the critics of bag bans,
see Adam Minter, "Stop Banning Plastic Bags, Please," Bloomberg, November 8, 2017,
www.bloomberg.com/opinion/articles/2017-11-08/stop-banning-plastic-bags
-please; Tanya Gold, "This Plastic Bag Conspiracy Is a Truly Deadly Distraction,"
Guardian, August 3, 2012, www.theguardian.com/commentisfree/2012/aug/03/plas
tic-bag-conspiracy-deadly-distraction.

39. Anderson, "To Ban or Not to Ban."

40. In the interest of full disclosure, I served as a subconsultant as part of this planning process.

Index

pliant wasteway, 106; and Boston's Zero Waste planning process, 141; and Boston waste composition study, 72; and curbside recycling programs, 108; and evolution of Seattle's defiant wasteway, 127; and evolution of weak recycling waste regime, 20–21, 22, 26, 42; experiential knowledge, 12, 14; expertise-as-process framework, 14; and framing of Boston's waste problem, 50, 52–63, 73–74; in Houston, 134; and redefinition of Seattle's waste, 85, 94; in the Sanitary City, 20, 22; and Seattle's planning and policy process, 85–86, 92, 95; and Seattle's transfer stations, 118–119; and support for WTE facilities, 50; and wasteways analytical framework, 11–16, 134; and weak recycling waste regime, 9

exporting garbage, 18

extended producer responsibility (EPR) legislation, 42, 133, 139–141

externality problem, 73

externalization of production costs, 7

extraction-manufacturing-consumption-waste chain: and American consumer culture, 6; and Boston's compliant wasteway, 103; and scope of waste problem, 2–3, 6; and Seattle's defiant wasteway, 130, 132; unequal impacts of, 7–8; and waste regimes concept, 4, 5; and wasteways analytical framework, 135; and the weak recycling waste regime, 5–6, 34, 43

Fairbanks, Gerard, 95

Fair Share Massachusetts, 62

federal government, 8–9, 65, 181n36. *See also specific federal institutions*

fee-for-service models, 57–60, 77, 85, 117–118, 137, 139. *See also* pay-as-you-throw (PAYT)

fines, 1, 125

Flynn, Raymond: and background of Boston's waste problem, 48–51; and parallel planning for Boston waste

system, 58–63; and politics of disposal problem, 63–70; and recycling programs, 105–106, 108, 165n58; and response to environmental impact report, 53; and stability of disposal problem, 71, 72–73

food waste, 6–7, 19, 100, 114, 116, 123–124. *See also* composting; organics and organic waste

Fox News, 126

framing of waste problem: and background of Boston's waste crisis, 49, 51–52; and background of Seattle's waste crisis, 78–79; and citizen input on Seattle's waste system, 92; and definition of waste in Boston, 73–74; and evolution of weak recycling waste regime, 19, 22, 27, 30–33, 35; and framing of Seattle's waste problem, 75–77; and parallel planning in Boston, 58, 62–63; and persistence of Boston's waste problem, 71–72; and political feuds, 69–70; and privacy issues, 125; and redefinition of Seattle's waste, 94; and regional waste planning, 60; and resource recovery in Seattle, 90–91; and stresses on weak recycling waste regime, 138–139; and wasteways analytical framework, 9–18. *See also* problem framing

Friedmann, John, 14

fuel, 7

funding for waste disposal, 108–109

Gale, Diana, 85, 95–96

Galle, Virginia, 85, *88*, 91

garbage scows, 32

Garbarino, Joe, 91

Gardner Street landfill, 46–47

Georgenes, Nicholas, 54

geospatial elements of garbage, 81

germ theory, 20

Gille, Zsuzsa, 3–6, 9, 15–16, 22, 41, 81

Gojenola, Bobbie, 87

Goldsmith, Amy, 106

Great Depression, 22

a

White, Kevin, 47–49, 52, 61
White, P. R., 37
Winthrop, Massachusetts, 66
women reformers, 19
World War II, 19, 22–23
Wynne, Brian, 13

yard waste composting, 77–78, 90, 100, 116, 124–125, 179n17
Young, Oran, 3

"Zero Waste Resolution," 112, 118, 141